Nuclear Data

A primer

IOP Series in Nuclear Spectroscopy and Nuclear Structure

Series Editors

John Wood, Georgia Institute of Technology, USA
Kristiaan Heyde, Ghent University, Belgium
David Jenkins, University of York, UK

About the Series:

The IOP Series in Nuclear Spectroscopy and Nuclear Structure provides up-to-date coverage of theoretical and experimental topics in all areas of modern nuclear spectroscopy and structure. Books in the series range from student primers, graduate textbooks, research monographs and practical guides to meet the needs of students and scientists.

All aspects of nuclear spectroscopy and structure research are included, for example:

- Nuclear systematics and data
- Ab-initio models
- Shell model-based descriptions
- Nuclear collective models
- Nuclear symmetries and algebraic descriptions
- Many-body aspects of nuclear structure
- Quantum mechanics for nuclear structure study
- Spectroscopy with gamma-rays, charged particles and neutrons following radioactive decay and reactions
- Spectroscopy of rare isotopes
- Detectors employed for gamma-ray, neutron and charged-particle detection in nuclear spectroscopy and related societal applications.

Nuclear Data

A primer

David G Jenkins
Department of Physics, University of York, Heslington, York, UK

John L Wood
School of Physics, Georgia Institute of Technology, Atlanta, Georgia, USA

IOP Publishing, Bristol, UK

Multimedia content is available for this book from http://iopscience.iop.org/book/978-0-7503-2674-2.

ISBN 978-0-7503-2674-2 (ebook)
ISBN 978-0-7503-2672-8 (print)
ISBN 978-0-7503-2675-9 (myPrint)
ISBN 978-0-7503-2673-5 (mobi)

DOI 10.1088/978-0-7503-2674-2

Version: 20210801

IOP ebooks

British Library Cataloguing-in-Publication Data: A catalogue record for this book is available from the British Library.

Published by IOP Publishing, wholly owned by The Institute of Physics, London

IOP Publishing, Temple Circus, Temple Way, Bristol, BS1 6HG, UK

US Office: IOP Publishing, Inc., 190 North Independence Mall West, Suite 601, Philadelphia, PA 19106, USA

Contents

Preface

Nuclear data represent information on a fundamental level of organization of matter, a unique manifestation of the quantum many-body problem, and a critical base for many technological applications. As a data base, it is vast and complex. Herein, we provide an introduction to the major features of nuclear data as an information resource, substantially independent of the methods by which the data were obtained or of physical and technological implications. Nevertheless, some basic quantum mechanics is needed to organize the data, some pointers to features important for applications are made, and essential details of the measurements behind the data are given.

Nuclei are the core of atoms. We are unaware of nuclei in the everyday world because atomic electrons 'shield' nuclei. Indeed, some familiarity with atomic data comes hand-in-hand with the process of organizing and understanding nuclear data. There is reciprocity in this understanding in that atoms only exist by virtue of the existence of nuclei and so a leading quest in science has been: 'Where do the chemical elements come from?' This has been answered through the realization that all but hydrogen and some helium and traces of a few other light elements are produced in stars via nuclear processes. Moreover, we have learned that some of the chemical elements are produced by the explosive coalescence of neutron stars, as recently observed in GW170817.

The study of nuclear structure (as also atomic structure) is dominated by observing energy changes and is broadly termed nuclear spectroscopy (atomic spectroscopy). Nuclear structure and atomic structure are two independent levels of organization of matter. Remarkably, they are both quantum mechanical domains, i.e. one must have some familiarity with quantum mechanics to understand data pertaining to either domain. This reach of quantum mechanics extends into molecular structure. Indeed, the manifestation of quantization at all three levels (nuclear, atomic, molecular) provides a 'cross-fertilization' of paradigms of quantum phenomena, e.g. both nuclei and molecules exhibit quantized rotations (but atoms do not).

The study of levels of organization of matter also extends more deeply to subnuclear structure, i.e. the description of nucleon structure: the protons and neutrons of nuclei are made up of quarks. Yet again, quantum mechanics provides the description of matter at the subnucleon level of organization. Quarks (q) form baryon (qqq), antibaryon ($\bar{q}\bar{q}\bar{q}$), and meson ($q\bar{q}$) combinations. At present, this forms the deepest level of matter for which we have established a microstructure. Independently of the quarks, leptons also play a fundamental role in the manifestations of matter. Electrons are present in our every-day world; muons and tauons have fleeting existence in high-energy processes. Their partner neutrinos are everywhere, but very hard to detect. At the level of nuclear spectroscopy, some knowledge of this particle zoo is essential.

In all of the manifestations of matter in the foregoing, the entities are organized into finite numbers of particle 'building blocks'. As such, they obey the laws of quantum mechanics for finite systems. This dictates conservation of angular momentum. Thus, familiarity with the quantization of angular momentum and the concept of spin is essential almost at the outset for pursuing the study of any of these systems. Perhaps, most dramatically, quantization of angular momentum in atoms underlies all of chemistry at the level of the Periodic Table. The building up (aufbau) of configurations in many-electron atoms is understood via the Pauli principle applied to uniquely labeled orbitals using angular momentum and spin quantum numbers of independent-particle behavior of the electrons. A few simple added rules provide a far-reaching qualitative basis for a near-complete understanding of the Periodic Table: Hund's rule (which favors maximally aligned spins in an angular momentum subshell due to magnetic effects) and screening effects of inner shell electrons (due to the Coulomb repulsion between electrons) are key concepts in this process.

The structure of many-nucleon nuclei can be approached via an aufbau process. However, this is an extremely limited approach in that while nucleons in many-nucleon systems exhibit some organization as independent-particles they do not exhibit a 'Hund's rule' and they do not repel each other. These issues that distinguish many-nucleon systems from (finite) many-electron systems provides the starting point for assembling and interpreting nuclear data.

Nuclear data are substantially organized by 'mass-chains' or 'A-chains'. This approach compiles and evaluates data for nuclear isobars (isobar = same mass number, $A = Z + N$, Z = proton number, N = neutron number), and is carried out under the auspices of the Nuclear Data Group who produce the Evaluated Nuclear Structure Data File (ENSDF). This is supported by unevaluated data compilations through the Unevaluated Nuclear Data List (XUNDL). A-chain evaluations are published intermittently in Nuclear Data Sheets. This resource is critical to the entire task of maintaining a usable and accessible base for nuclear science. Unless otherwise stated, all data shown in the figures and tables are taken from these sources.

While compilation and evaluation of nuclear data are ordered by A chain (sometimes just an individual isotope), 'horizontal' compilations and evaluations are limited or rare. By horizontal is meant a specific nuclear property across the entire range of A, Z, and N, sometimes referred to as the 'mass surface'. A leading example is that of nuclear masses, maintained by the Atomic Mass Evaluation network. The present endeavor is horizontal and it aspires to a global perspective.

To encourage our readers to engage in the process of accessing and manipulating nuclear data, we embed a number of short video tutorials in the electronic version of this book (see exercises 1-23, 1-24, 2-11, 3-6 and 4-11). These are designed to demonstrate how the process of working with nuclear data is carried out in practice. The goal is to familiarize the reader with concepts and quantities that are 'derived' from the data sources, i.e. the concepts and quantities cannot be seen directly by just looking at the data. These tutorials range from basic extrapolation of the perspective available from established data bases to nuclei not yet studied, to the use of the equations for deduction of properties such as electromagnetic moments (which cannot be observed directly in nuclei). The examples chosen and the level of detail presented are aimed at providing the reader with the sense that they can exercise control of the handling of nuclear data and the equations and basic theory concepts involved. Ultimately, our goal is to impart a sense of the engagement in exploration that can be reached by becoming familiar with nuclear data.

Acknowledgements

Many colleagues have influenced our choice of material and emphasis in this volume; we particularly wish to thank Kris Heyde and Bob Wadsworth for a detailed reading and feedback on this work.

Author biographies

David G Jenkins

Professor David Jenkins is Head of the Nuclear Physics Group at the University of York, UK. He is also a Fellow of the Institute of Advanced Study, University of Strasbourg (USIAS) and an Extraordinary Professor of the University of Western Cape in South Africa. Professor Jenkins' DPhil was in experimental nuclear physics at the University of York. Following postdoctoral research at Argonne National Laboratory and the University of Liverpool, he returned to the University of York as an EPSRC Advanced Fellow and subsequently joined the academic faculty there. His research in experimental nuclear physics focuses on several topics such as nuclear astrophysics (the origin of the chemical elements), clustering in nuclei and the study of proton-rich nuclei. His research is truly international and he has led experiments at accelerator facilities across Europe, the US, Canada, Japan and South Africa. In recent years, the author has developed a strong strand of applications-related research with extensive industrial collaboration. He has led the development of bespoke radiation detectors for homeland security, nuclear decommissioning, bore-hole logging and medical applications.

John L Wood

John L Wood is a Professor Emeritus in the School of Physics at Georgia Institute of Technology. He continues to collaborate on research projects in both experimental and theoretical nuclear physics. Special research interests include nuclear shapes and systematics of nuclear structure.

Chapter 1

Gross properties of nuclei

Gross properties of nuclei—masses and radii—are introduced as systematics with respect to mass number, and proton and neutron number. The emphasis is that nuclear structure is the study of a quantum mechanical many-body problem. Local views mislead: a global view is essential. Differences in gross properties reveal fundamental aspects of nuclear structure: shell structure, pairing correlations and deformation. Parallels between the Chart of the Nuclides and the periodic table are illustrated, especially independent-particle motion in complex many-body systems. But, nuclei are very different: they possess many-body correlations and quadrupole deformation. This sets the stage for the entire book.

Concepts: chart of the nuclides, atomic mass, root-mean square charge radius, surface diffuseness, isotope shift, charge volume, binding energy, binding energy per nucleon, short-ranged forces, Coulomb force, pairing, shell structure, separation energy, quadrupole moment, drip line, beta decay, x-ray, energy barrier, superheavy element, independent-particle motion.

Nuclei are finite many-body quantum systems. As such it is necessary to accumulate nuclear data in a systematic manner over long chains of isotopes (changing N) and isotones (changing Z). Figure 1.1 shows the current extent of nuclei (combinations of Z and N) that have been characterized by at least one nuclear property.

The term 'gross properties' refers to quantification of properties that involves the entirety of the nucleus, i.e. of all of the nucleons (protons and neutrons) making up the given nucleus. In particular, the mass and the radius of a nucleus are understood to fall within this definition. At the outset, a subtle distinction must be made: a mass is usually measured for an atom, whereas a radius is measured for a nucleus.

The mass of an atom (nucleus plus complement of atomic electrons) is the preferred entity for a mass measurement because it is easier to handle than a bare nucleus. Producing a bare nucleus would incur stripping away all of the atomic electrons: this is difficult to do because the highly charged ion would easily be

doi:10.1088/978-0-7503-2674-2ch1

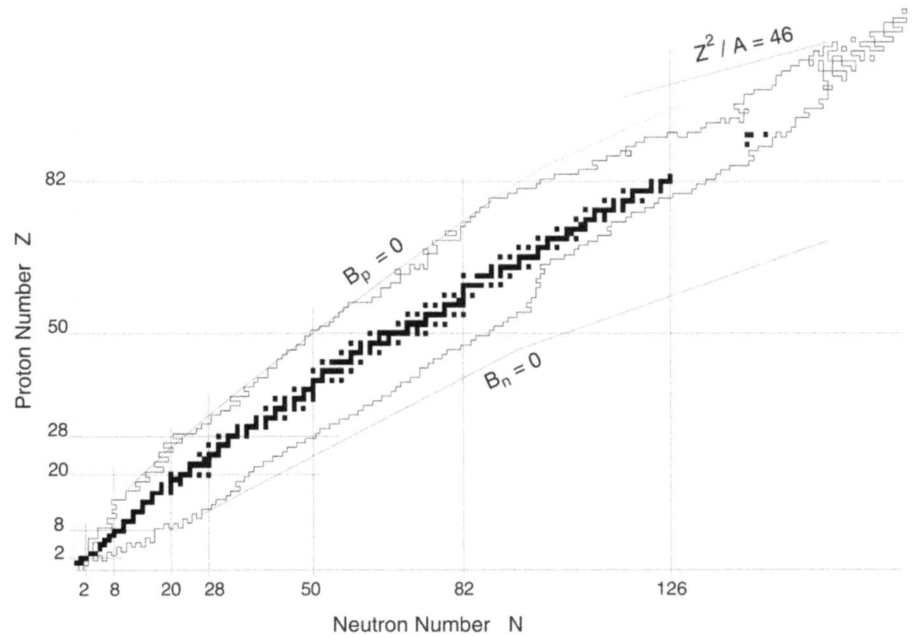

Figure 1.1. The known atomic nuclei are presented in a standard manner, termed the Chart of the Nuclides where 'nuclide' means nucleus plus full complement of atomic electrons. The term 'isotopes' is used to refer to the individual entries in the chart and also to refer to a sequence of constant Z. Sequences of constant N are referred to as 'isotones'. The black squares denote the isotopes found in Nature. The horizontal and vertical lines are historically called the 'magic numbers': they correspond to energy shell gaps observed in spherical nuclei at the nucleon numbers—2, 8, 20, 28, 50, 82, and (for neutrons) 126. The lines marked $B_p \approx 0$, $B_n \approx 0$ are the so-called proton and neutron 'drip' lines: outside of these borders, it is estimated that nucleons will be lost so rapidly that the species cannot be isolated for study. (See the video-based exercises, exercises 1-23 and 1-24 to explore this issue further.) The line marked $Z^2/A = 46$ is an estimate of the limit of nuclei that can be isolated for study with respect to limits due to rapid spontaneous fission. The stepped border is the current (ca. 2010) limit for which at least one characteristic of an individual isotope has been characterized. Reproduced from [1], copyright 2010 World Scientific Publishing Company.

neutralized by capturing electrons from its environment and so such a species is not stable. Usually for mass measurements, singly charged ions are used because they are easy to maintain and manipulate using electric and magnetic fields (and a good vacuum).

The 'radius' of a nucleus also has a subtlety in its definition. A nucleus does not possess a sharp surface—it exhibits a diffuseness as deduced from inferred charge density distributions determined by elastic electron scattering[1]:

$$\rho(r) = \rho_0 \left\{ 1 + \exp\left(\frac{r - R_B}{a} \right) \right\}^{-1}. \tag{1.1}$$

[1] Note that such models of nuclear charge distributions are also explored using muonic atom x-ray spectroscopy. Muons 'orbit' much closer to the nucleus than electrons and thus provide further information on $\rho(r)$. Details are given in chapter 6.

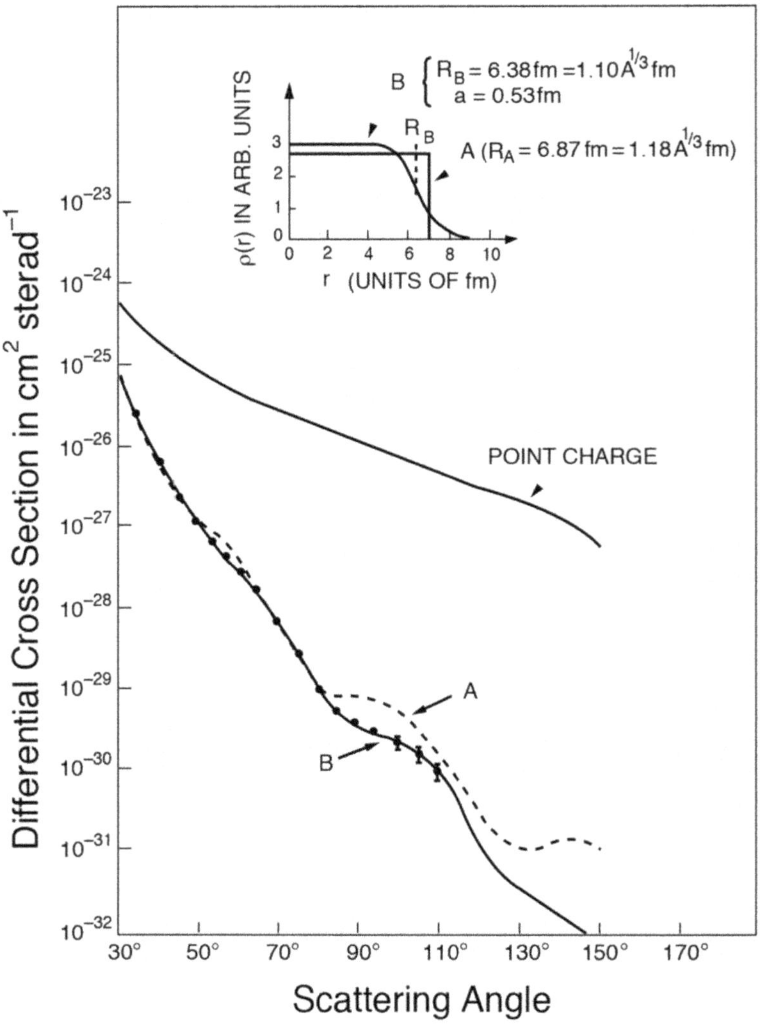

Figure 1.2. Elastic scattering of electrons as a function of angle with respect to the electron beam for the isotope ^{197}Au reveals, via the fitting of a model charge-density distribution, $\rho(r)$ (see text), that nuclei do not have sharp surfaces. The inset shows two model choices: a constant density with a sharp surface (A), radius R_A and a diffuse surface (B), cf. equation (1.1). The pattern for scattering from charge density distributions A, B, and a point charge is indicated. The electron energy, 153 MeV, corresponds to a de Broglie wavelength of 8 fm. Reproduced from [1], copyright 2010 World Scientific Publishing Company. The original data presented in this figure are from [2–4].

This is illustrated in figures 1.2 and 1.3. Nuclear surface diffuseness shows systematic trends but is not a fixed geometric property of nuclei. In consequence it is convenient to define a root-mean-square charge radius ($\langle r^2\rangle^{1/2}$) for a given nuclear species, i.e. a given proton number (Z) and neutron number (N). This is well-defined and is unique for a given (Z, N). Such data are usually shown for fixed Z as a function of N and,

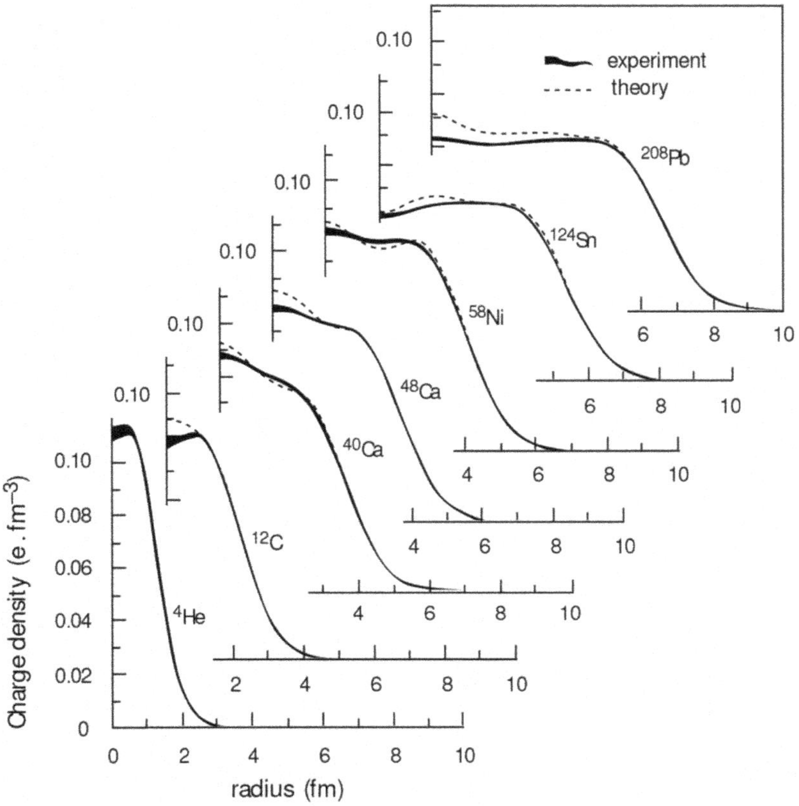

Figure 1.3. Ground-state charge density distributions for selected nuclei. These were determined in a manner similar to that depicted in figure 1.2. The dashed curves are theoretical estimates based on mean-field theory. The thickness of the lines represents the experimental uncertainty. Note the central densities steadily decrease, from ~0.11 $e \cdot \mathrm{fm}^{-3}$ (^{4}He) to ~0.06 $e \cdot \mathrm{fm}^{-3}$ (^{208}Pb), with increasing mass because of the increasing neutron excess. Reproduced from [1], copyright 2010 World Scientific Publishing Company. The original data leading to this figure are found in [5].

further, what is plotted is the difference between $\langle r^2 \rangle$ with reference to a specified N value; they are called *isotope shifts*.

An example of a sequence of isotope shifts is shown in figure 1.4, where the data were obtained using optical hyperfine spectroscopy. A scheme that illustrates the essential physics of such spectroscopy is presented in figure 1.5. At the present level of detail, the nucleus can be regarded as a small but finite 'volume' of positive charge at the center of an atom and the atomic electrons respond to this charge volume in a small but precisely quantifiable manner measured as $\langle r^2 \rangle$ using tunable lasers.

Nuclear masses could be naively thought to be given by $M(Z, N) = Zm_p + Nm_n$, where m_p is the mass of the proton and m_n is the mass of the neutron. However, a given nucleus (Z, N) possesses a well-defined binding energy and this is on the order of 1% of the mass (recall $E = mc^2$) of the total nucleus. Thus, the mass of a nucleus is less than that of its constituent parts. Atomic masses include Zm_e minus the total

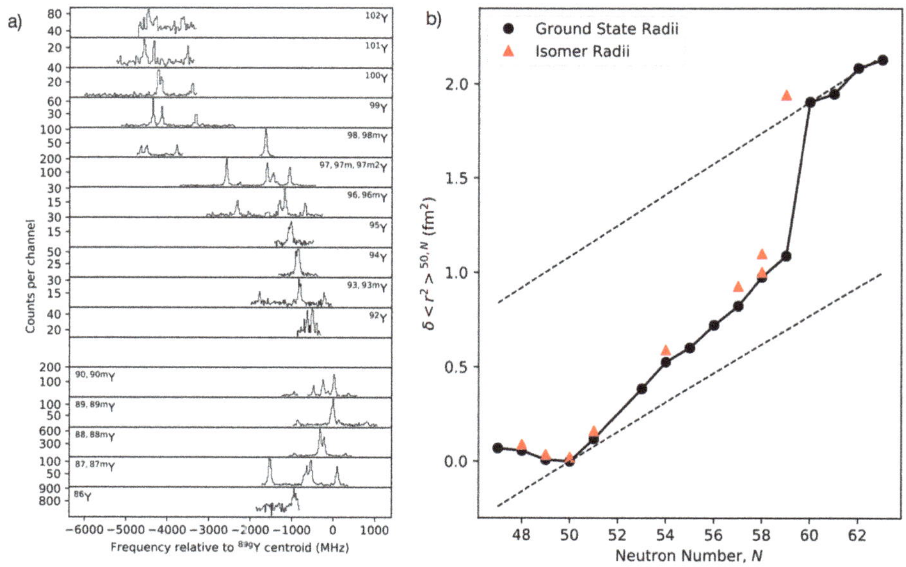

Figure 1.4. Isotope shift data for the yttrium ($Z = 39$) isotopes. In (a) the hyperfine spectrum for each isotope is shown. In (b) the deduced isotope shifts relative to ^{89}Y ($N = 50$) are shown. Note that the 'centroid' of each hyperfine spectrum shifts in a way that correlates with the isotope shift. The dashed lines are model estimates of the isotope shifts, with reference to ^{89}Y, and the details are beyond the present discussion; however, see exercise 1-3. (Reprinted with permission from [6]. Copyright (2007) by Elsevier.)

electronic binding energy. But the electronic binding energy is usually less than the measurement uncertainty[2] of the total atomic mass. Therefore, we discuss

$$M(Z, N) = Zm_p + Nm_n + Zm_e - B(Z, N), \tag{1.2}$$

where $B(Z, N)$ is the nuclear binding energy. Table 1.1 shows selected values of $M(Z, N)$ and $B(Z, N)$. While tabulation of binding energies does not look especially interesting, the quantity B/A, $A = Z + N$, the *binding energy per nucleon* is most interesting. Figure 1.6 shows B/A for selected nuclei. The following features should be noted:

1. For the majority of nuclei, $B/A \sim 8.3$ MeV/nucleon, with a variation of ±5% for $20 < A < 200$.
2. The maximum value of B/A occurs for ^{62}Ni.
3. There is a steep decline in B/A going to the lightest nuclei (except for ^{4}He, cf. table 1.1).
4. There is a steady and slowly steepening decline of B/A towards heavy nuclei.

The following interpretations can be made:

1. The force that binds nucleons must be short ranged. A short-ranged force predominantly only influences nearest neighbors with the result that there is a

[2] Atomic binding energies can be deduced by summing the ionization potentials for successive removal of the electrons in a given atom.

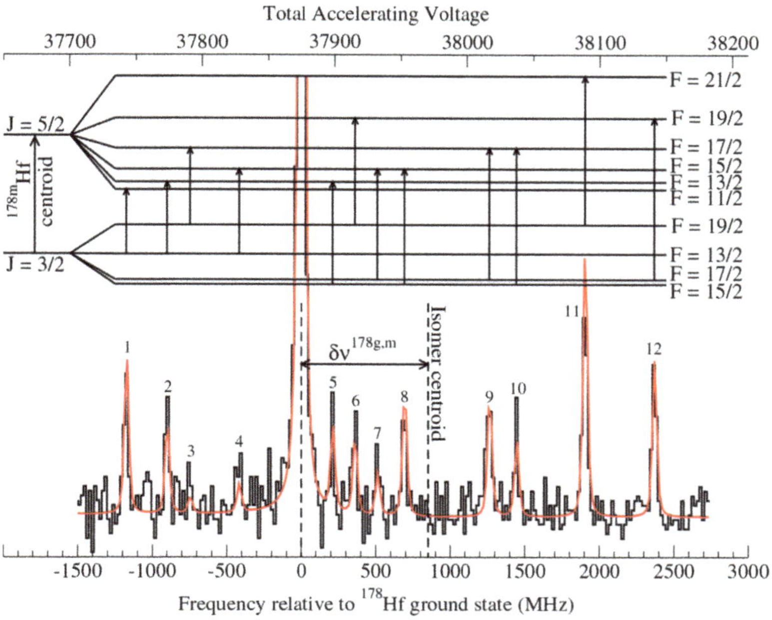

Figure 1.5. Atomic hyperfine structure for the nucleus ^{178}Hf. The atomic transition is shown on the left and the hyperfine structure due to a $I = 8$ excited state (^{178m}Hf, $T_{1/2} = 4.0$ s), relative to the $I = 0$ ground state (no hyperfine structure) is shown on the right. The hyperfine spin ordering for $I = 8$ is dominated by the magnetic moment of the state and is monotonic, but there is a reordering of the spins for the lower multiplet due to the quadrupole moment of the $I = 8$ state. The total accelerating voltage refers to the fact that ions were accelerated to velocities where they were brought into resonance with a fixed laser beam frequency by 'Doppler shifting'. This avoided the need to retune the frequency of the laser beam. (Reprinted from [7] with permission of IOP Publishing.)

'saturation' of the binding. This is a characteristic property of a liquid, e.g. water exhibits a constant boiling point (latent heat of vaporization) that is independent of the quantity of water being heated. A long-ranged attractive force would exhibit a non-linear and increasing binding energy. A long-ranged repulsive force would exhibit a non-linear and decreasing binding energy: this is manifested in heavy nuclei because of the increasing number of protons and their long-ranged Coulomb repulsion.

2. The steep decline in B/A for light nuclei is also a manifestation of the short range of the nucleon–nucleon interaction: many nucleons in light nuclei are at the nuclear surface and so they do not have a full complement of neighbors such as occurs for most nucleons in a heavy nucleus. The binding of light nuclei can be described as 'unsaturated'.
3. The maximum value for B/A is a balance between minimum surface area-to-volume ratio and minimum Coulomb repulsion. While the occurrence at ^{62}Ni is most interesting—it is near where stellar fusion comes to a halt (and a supernova can occur)—it is not a profound property of nuclei.

The liquid-drop-like behavior of nuclei is also reflected in $\langle r^2\rangle^{1/2}$ values when plotted as a function of A, viz.

Table 1.1. Masses and binding energies for selected nuclei, given in MeV. Note that the atomic mass unit u = 931.494 102 42(28) MeV/c^2 = 1.660 539 066 60(50) × 10^{-27} kg. Values are taken from AME2016 [8]. Note, since the completion of this book, a new mass evaluation has been completed, AME2020, and there are some small changes to some of these numbers.

Isotope	$M(A, Z, N)$	$B(A, Z, N)$
n	939.57	0
^{1}H	938.78	0
^{2}H	1 876.12	2.22
^{4}He	3 728.40	28.30
^{5}He	4 668.70	27.56
^{6}He	5 606.56	29.27
^{8}Be	7 456.89	56.50
^{12}C	11 177.93	92.16
^{62}Ni	57 685.89	545.26
^{96}Zr	89 337.99	829.00
^{96}Nb	89 337.83	828.38
^{96}Mo	89 334.64	830.78
^{100}Zr	93 073.03	852.22
^{138}Te	128 480.49	1138.86
^{238}U	221 742.91	1801.69

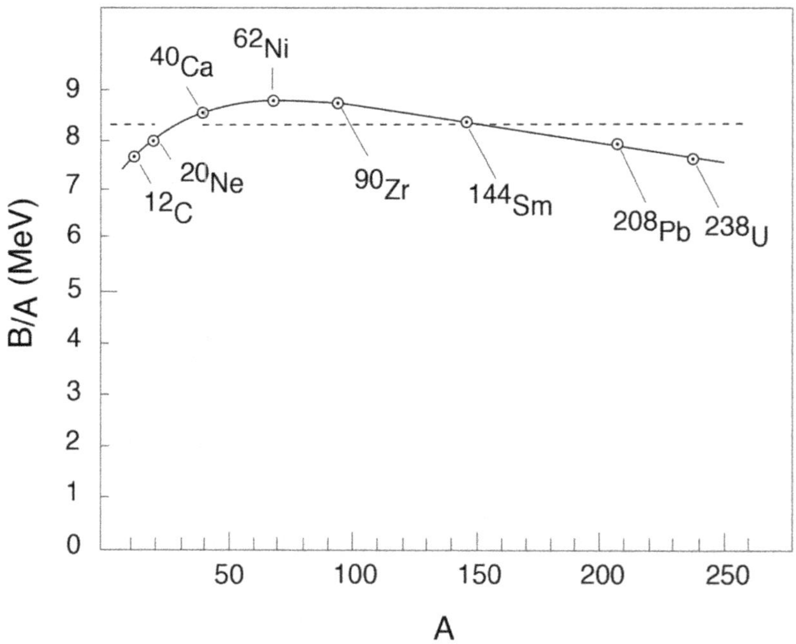

Figure 1.6. Binding energy per nucleon. The detailed features are discussed in the text. The solid curve is a smooth line through the data. The data points are chosen arbitrarily. The data are taken from AMDC files (see chapter 2, section 3). Reproduced from [1], copyright 2010 World Scientific Publishing Company.

$$\left[\frac{5}{3}\langle r^2\rangle\right]^{1/2} = R_0 A^{1/3}. \tag{1.3}$$

Figure 1.7 shows such a plot and one observes that the volume of a nucleus scales as A, the number of nucleons. If the nuclear force were long ranged and attractive, nuclear volumes would 'shrink' relative to the number of particles. This occurs for atoms (as shown in figure 1.8) and reflects the dominance of the long-ranged attraction of the $+Ze$ charge of the nucleus over the electrons as Z increases; there are secondary electron–electron repulsive forces, but these are less important due to the diffuse distribution of the electrons within the atom. Figure 1.8 also shows the manifestation of atomic shell structure in atomic radii.

Differences in gross properties of nuclei reveal underlying structure. Figure 1.9 shows the manifestation of nuclear shell structure in differences for nuclear radii. Figure 1.10 shows differences in nuclear binding energies expressed as separation energies, e.g. one-neutron separation energies, S_n

$$S_n = -M(A, Z, N) + M(A - 1, Z, N - 1) + m_n \tag{1.4}$$

Three features should be noted in figure 1.10:

1. There is an odd–even 'staggering' of S_n.
2. There are discontinuities in the form of 'steps' down with increasing N or A.
3. The trend between the steps is smooth and down-sloping with increasing neutron number.

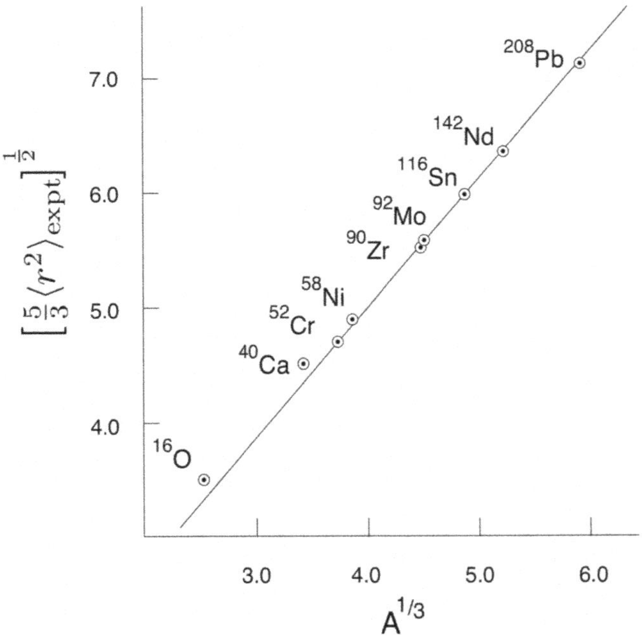

Figure 1.7. Nuclear root-mean-square charge radii relative to the liquid-drop model. The solid line is a best (straight-line) fit for nuclei with $A \geqslant 90$. The units are femtometers (1 fm = 10^{-15} m). The deviation of the lighter nuclei from this line is due to surface diffuseness effects (cf. figure 1.3). The data are taken from [9]. Figure reproduced from [1], copyright 2010 World Scientific Publishing Company.

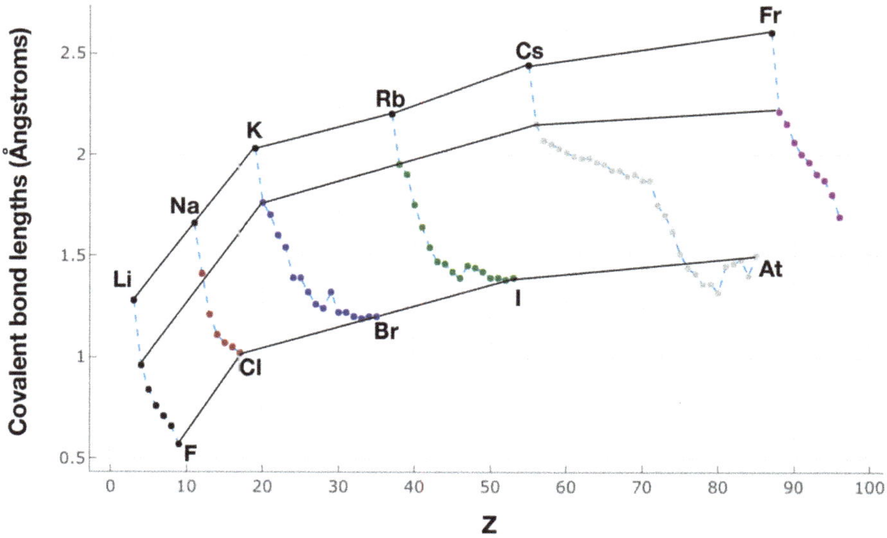

Figure 1.8. Atomic radii expressed as covalent bond radii. Note the contraction that occurs successively through each sequence Li–F, Na–Cl, etc. This reflects the long-ranged nature of the Coulomb attraction of the nucleus as the atomic number increases. Other features are due to electron–electron repulsion and electronic subshell structure. Note, the use of the Ångstrom = 0.1 nm.

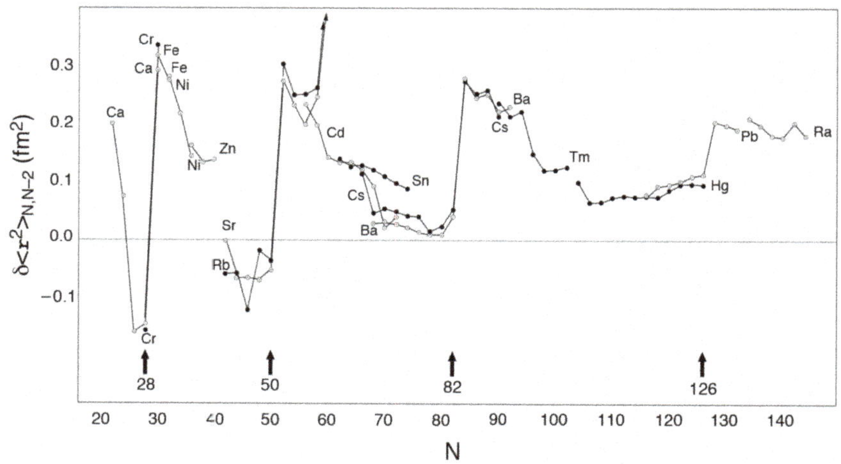

Figure 1.9. Isotope shifts for selected isotopic sequences of nuclei. The data are taken from [9]. The figure is reproduced from [1], copyright 2010 World Scientific Publishing Company.

The following interpretations can be made:

1. The odd–even staggering is due to enhanced binding when the neutron number is even. (A similar effect is observed for S_p.) This effect is termed 'pairing'. This is a profound manifestation of quantum mechanical correlations. Details will be addressed in later chapters.
2. The steps are due to so-called 'shell' closures. The effect is clearer when S_{2n} is plotted, as shown in figure 1.11. Proceeding from right to left in the figure,

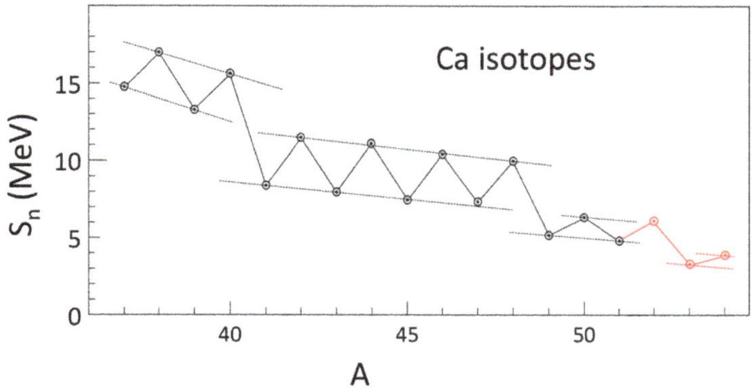

Figure 1.10. Nuclear one-neutron separation energies, S_n. The data are taken from AMDC files (see chapter 2, section 2.3). Reproduced from [1], copyright 2010 World Scientific Publishing Company.

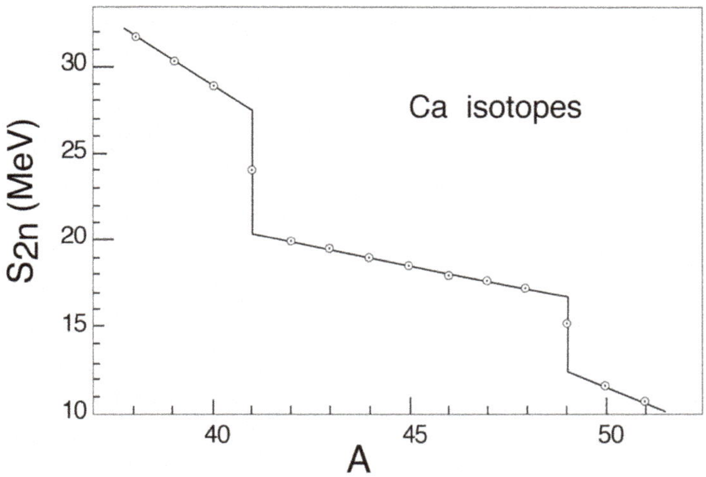

Figure 1.11. Nuclear two-neutron separation energies, S_{2n}. The data are taken from AMDC files (see chapter 2, section 2.3). Reproduced from [1], copyright 2010 World Scientific Publishing Company.

successive neutrons (neutron pairs) are being removed. Nucleons in the nucleus are confined in an average potential generated by all of the other nucleons. There is an energy ordering and a sequential filling of orbitals with well-defined occupancies of orbitals which possess degeneracies. This is shown schematically in figure 1.12. When an energy gap is reached in the removal process, there is a sudden step up in the removal energy. This is because removal is progressively from deeper-lying orbitals in the potential and so more energy must be supplied to effect removal.

3. The trend between shells has a shallow slope because the size of the confining potential changes as $\sim A^{1/3}$ (cf. equation (1.3). Thus, with increasing A, the width of the potential increases and the potential is less deep.
4. A global view of neutron shell closures is provided by S_{2n}, as shown in figure 1.13.

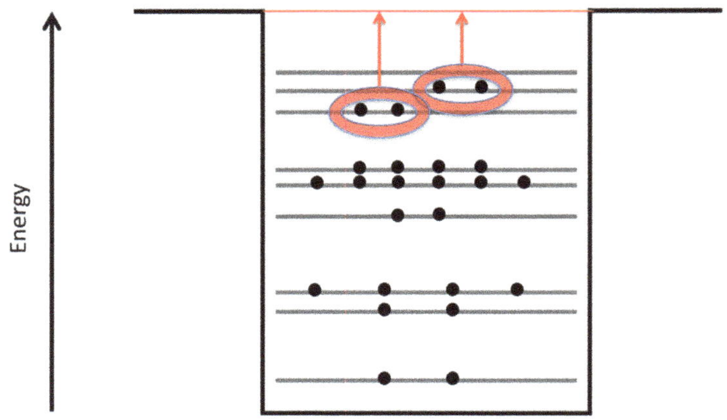

Figure 1.12. A schematic view of level filling and nucleon (pair) removal in a nucleus. There are separate sets of levels for protons and neutrons. The arrows depict removal of successive pairs from the nucleus, cf. figure 1.11 (note, this occurs from right to left).

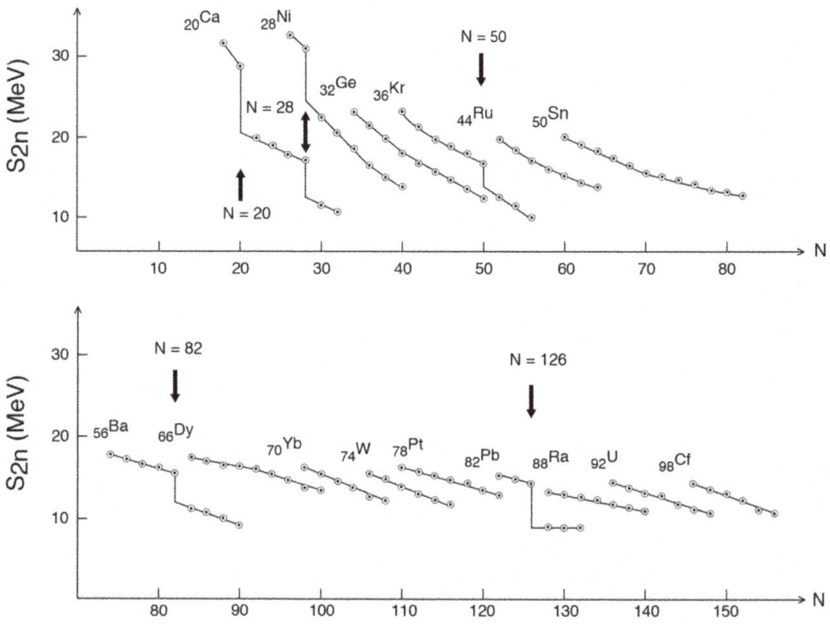

Figure 1.13. Two-neutron separation energies, S_{2n}. Note, the shells occur at *and only at* $N = 20$, 28, 50, 82, and 126. The data are taken from AMDC files (see chapter 2, section 2.3). Reproduced from [1], copyright 2010 World Scientific Publishing Company.

The *shell structure* of nuclei was the first key step towards organizing nuclear data using quantum mechanics and was formalized in 1949 in two papers, one by Maria Goeppert-Mayer [10] and one by Haxel *et al.* [11]. For this insight, Goeppert-Mayer and Jensen shared the 1963 Physics Nobel Prize (with Eugene Wigner). The shell structure of nuclei led to the *nuclear shell model*, details of which are given in later chapters.

One might suppose that nuclei, in their apparent conformity to liquid drop-like behavior, are always spherical. A spherical shape would minimize the surface area and so maximize the binding energy by maximizing the saturation of the short-ranged nucleon–nucleon force. However, most nuclei are deformed. The evidence for this comes from *quadrupole moment* measurements. Quadrupole moments of selected nuclei are shown in figure 1.14. This aspect of nuclear behavior will dominate much of this book. It is another profound manifestation of quantum mechanical correlations. Details will be addressed later. However, it can be noted that the first evidence for nuclear deformation came from optical hyperfine spectroscopy conducted by Schüler and Schmidt [12], as interpreted by Casimir [13]. The saga of nuclear deformation is detailed by Heyde and Wood [14] (and further details of the early history are given by Lieb [15]).

There are limits to the existence of nuclei with respect to the possible combinations of Z and N that survive long enough to be isolated for study in the laboratory. An example of these limits is illustrated in figure 1.15. Thus, we speak of proton and neutron 'drip lines'—borders beyond which nuclei do not exhibit bound states. However, such limits depend upon the environment of the nucleus in that stellar interiors provide an environment where nuclei that cannot be isolated in the laboratory may have a fleeting existence sufficient for the species to undergo a

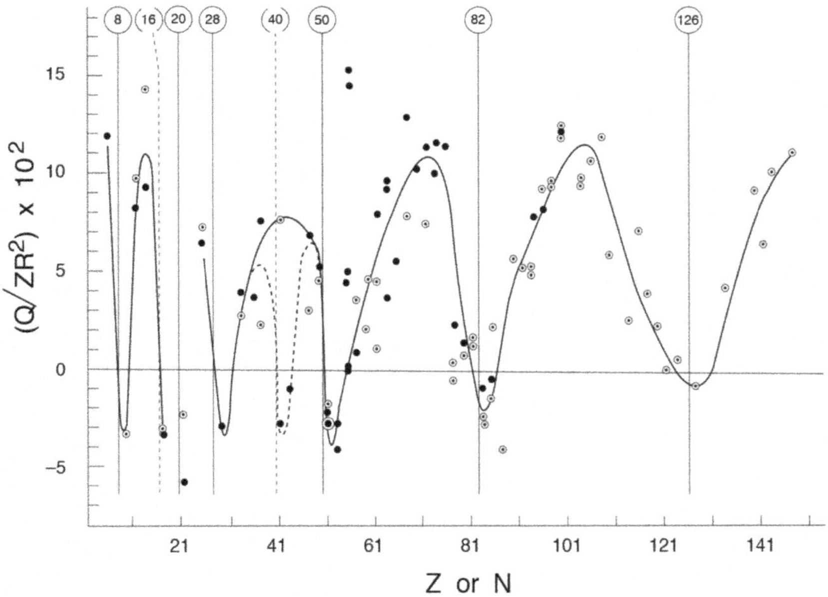

Figure 1.14. Electric quadrupole moments indicate that many nuclei are non-spherical, both for odd Z and odd N. The symbol code is odd-Z (solid), odd-N (open). Note that positive values (prolate shapes = rugby football-like) dominate over negative values (oblate shapes = doorknob-like): there is not a consensus view on why this occurs. The closed shell nucleon numbers are distinguished by values that are consistent with zero. The normalization, by ZR^2 is to accommodate the dimensions of the quantity being plotted so that the underlying nuclear deformation can be compared. The figure is reproduced from [1], copyright 2010 World Scientific Publishing Company. The original data that are plotted in this figure are taken from [16].

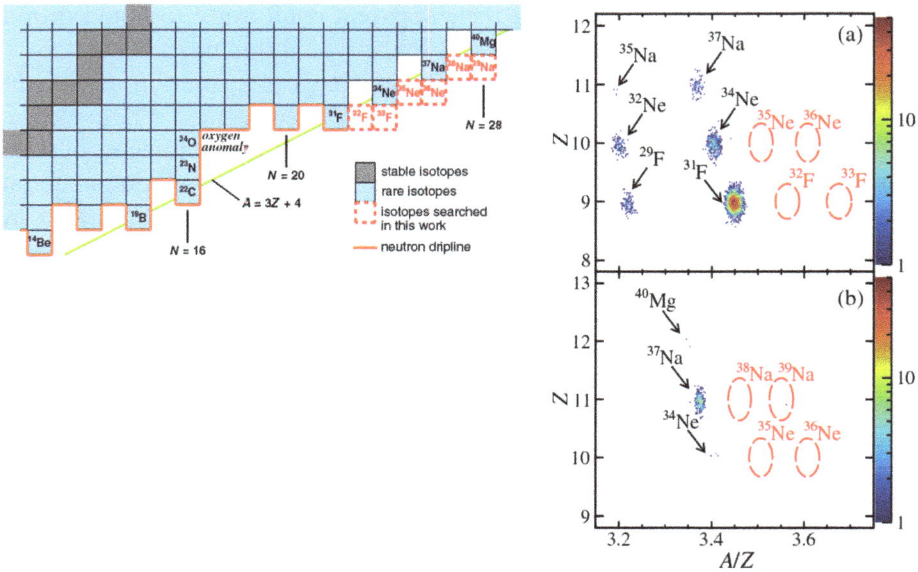

Figure 1.15. The neutron 'drip line' reached for fluorine, neon, and sodium nuclei. (Reprinted with permission from [17]. Copyright (2019) by the American Physical Society.)

nuclear process. The most important example of this is the nucleus ^{8}Be, which exists long enough in massive stars to lead to the formation of carbon (^{8}Be + ^{4}He → ^{12}C). Some details of this and the nuclear structure involved are given later in the book. The extraordinary thing is that essentially all of the carbon and heavier elements in the Universe were synthesized in stellar interiors via this carbon-producing process.

For a given $A = Z + N$, a plot of nuclear masses versus Z reveals that there is a parabolic behavior, as shown in figure 1.16 and 1.17. These plots show that there is a most-stable value of Z for a given A. The consequence of this is the family of nuclear decay processes called *β decay*. Nuclei that lie on the left-hand side of the parabolas undergo β^- decay, whereby a neutron becomes a proton with the radiative emission of an electron (β^- particle) and an electron antineutrino, $\bar{\nu}_e$. Nuclei that lie on the right-hand side of the parabolas undergo β^+ decay, whereby a proton becomes a neutron with the radiative emission of a positron (β^+ particle) and an electron neutrino, ν_e. This process can also occur by *electron capture decay* (a bound atomic electron plus a proton can undergo a process whereby they become a neutron with the radiative emission[3] of an electron neutrino). There is also the possibility for certain even–even nuclei to undergo *double-beta decay*. This is indicated in figure 1.17. Some details of beta decay and electron capture are given in later

[3] Electron capture is usually associated with the emission of *x-rays*. This is because the captured atomic electron leaves a 'vacancy' in the atomic shell (most probably the K shell) from which it was captured. A *K vacancy* is 'filled' by the transition of an L-shell (or M-shell,) electron with the emission of K x-rays. The process continues until the atomic electrons have 'relaxed' into a completely ordered filling of the shells. Recall, shells in many electron atoms are labeled K, L, M, N, O, P. Note, there is a competing process of vacancy filling, the so-called 'Auger process'. A few details are given in chapter 6, see figure 6.12.

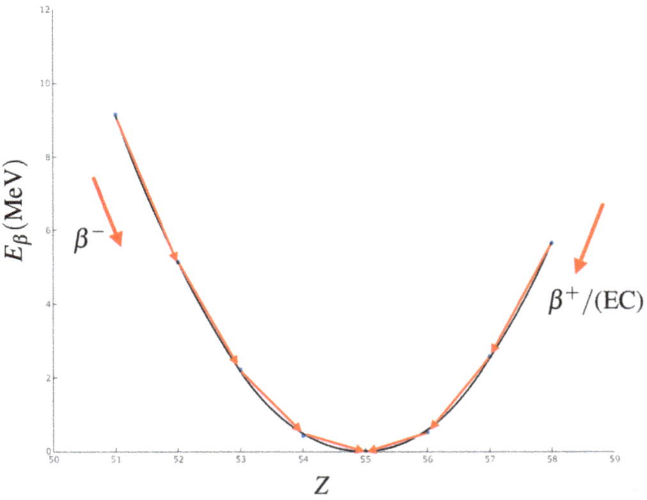

Figure 1.16. The A = 133 mass parabola. Masses are shown relative to $^{133}_{55}$Cs, which is stable. The relative energies are the energies available for β decay. The arrows show the β decays that occur. Note: the decay energy for ^{133}Ba → ^{133}Cs is 517 keV and so it can only take place by electron capture (β^+ decay requires a decay energy >1022 keV). The data are taken from AMDC files (see chapter 2, section 2.3).

chapters. The minimum in these parabolas defines the so-called line of stability in the chart of the nuclides.

Nuclear masses reveal another property of nuclei that is usually hidden, namely that a variety of decay processes besides β decay are energetically favorable. These processes include α decay, spontaneous fission, proton decay and emission of heavy nuclear clusters such as ^{14}C. If the mass relation

$$M(A, Z, N) > M(A - 4, Z - 2, N - 2) + M(^4\text{He}) \tag{1.5}$$

holds, it is energetically favorable for an α particle (^{4}He nucleus) to be emitted from the nucleus characterized by (A, Z, N). However, for this to happen the α particle has to 'tunnel' through a 'Coulomb barrier'. The process is depicted in figure 1.18. It is a fundamental quantum mechanical process. Coulomb barriers exist for all such charged-particle decay processes of nuclei. For many nuclei these processes are energetically favorable, but are too improbable for observation to detect the process. Some examples of half-lives for such processes are given in table 1.2.

There is one direction in which combinations of Z and N are still being explored with no known limitations except the difficulty of making them in the laboratory: the *superheavy elements*. Currently, the farthest reach is to element 118, for which evidence of Z = 118, N = 176 has been obtained. This was achieved by bombarding a ^{249}Cf target with a beam of ^{48}Ca nuclei, with 'evaporation' of three neutrons. Element 118 has been named Oganesson in honor of Yuri Oganessian who has been a major pioneer of superheavy element exploration [18]. Element 118 is believed to be a chemical homolog of the inert gases and so follows on from xenon (Z = 54) and radon (Z = 86), hence the name ending. Implicit in this book is the presence of Dimitri Mendeleyev's periodic table of the chemical elements, shown in figure 1.19.

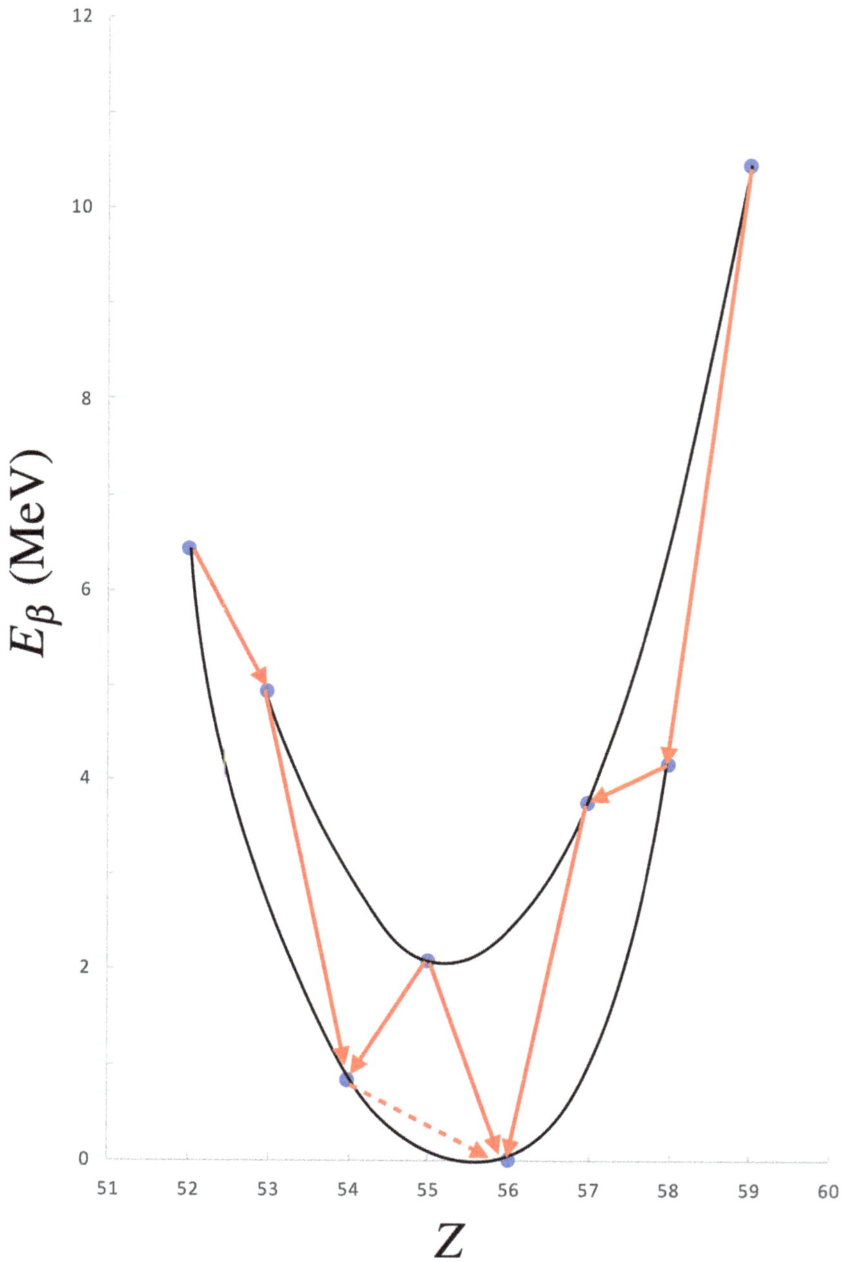

Figure 1.17. The A = 134 mass parabolas. Two parabolas occur for even masses because of an attractive pairing interaction between like nucleons (see text). Thus, $^{134}_{55}Cs$ undergoes β^+ decay to $^{134}_{54}Xe$ and β^- decay to $^{134}_{56}Ba$. The dashed arrow indicates double beta decay: the half-life for this process is $>10^{18}$ y and has not yet been studied for ^{134}Xe. The data are taken from AMDC files (see chapter 2, section 2.3).

It can be conjectured that the limit of the periodic table will be determined by Coulomb repulsion of the protons and the likely near instant spontaneous fission of the species. The half-life of the species will be dictated by the Coulomb barrier through which the separating pair of fission fragments must tunnel. The most

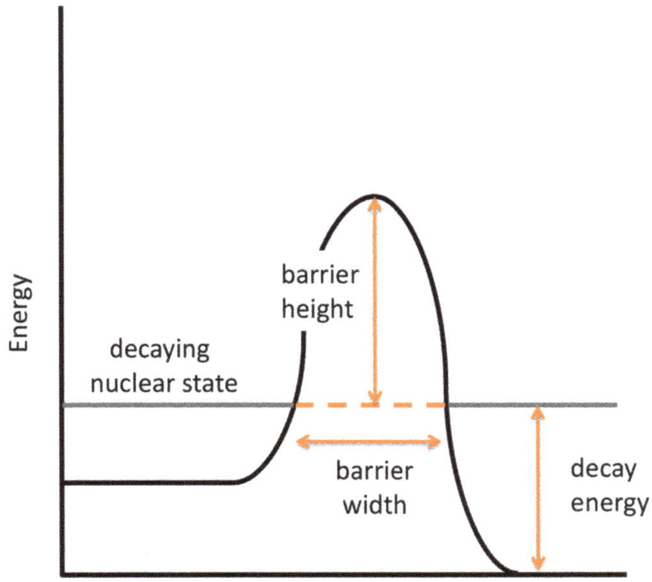

Figure 1.18. Schematic view of a Coulomb barrier for charged-particle (nucleon cluster) emission from a nucleus undergoing such a decay. The probability of decay is determined by the barrier height and width and the mass of the cluster. This is the process of quantum mechanical tunneling.

Table 1.2. Decay modes, energies, and half-lives for selected nuclei. Note the extreme ranges: these reflect the nature of the decay combined with structural effects such as spin changes. Note further, the half-lives quoted for the decay processes are sometimes partial half-lives, $T_{1/2}^{\mathrm{partial}} = T_{1/2}$ / decay branch fraction. These data were taken from ENSDF in 2020 when the data 'cutoff' was made for the book (Dec. 2020). Since that time, some of these numbers have changed.

Isotope	Decay mode	Decay energy (MeV)	$T_{1/2}$
^{8}Be	$\alpha + \alpha$	0.09	[5.57(25) eV]
^{12}Be	β	11.807	21.46(5) ms
^{96}Zr	β	0.164	>2.4 × 10^{19} y
^{96}Zr	$\beta\beta$	3.356	2.0(4) × 10^{19} y
^{113}Cd	β	0.324	8.04(5) × 10^{15} y
^{144}Nd	α	1.850	2.29(16) × 10^{15} y
^{151}Lu	p	1.233	127(4) ms [total: 80.6(20) ms]
^{212}Po	α	8.785	294.3(8) ns
^{224}Ra	^{14}C		2.5(6) × 10^{6} y [total: 3.631 9(23) d]
^{238}U	α	4.270	4.468(6) ×10^{9} y
^{238}U	fiss.	≈190	8.20(10) ×10^{16} y
^{252}Cf	fiss.		85.33(26) y [total: 2.647(3) y]

promising direction for synthesizing new superheavy elements would be to find a way to 'add' more neutrons.

The organization of the chemical elements reflects a shell structure that exists for electrons in many-electron atoms. This is already evident in figure 1.8 for atomic radii.

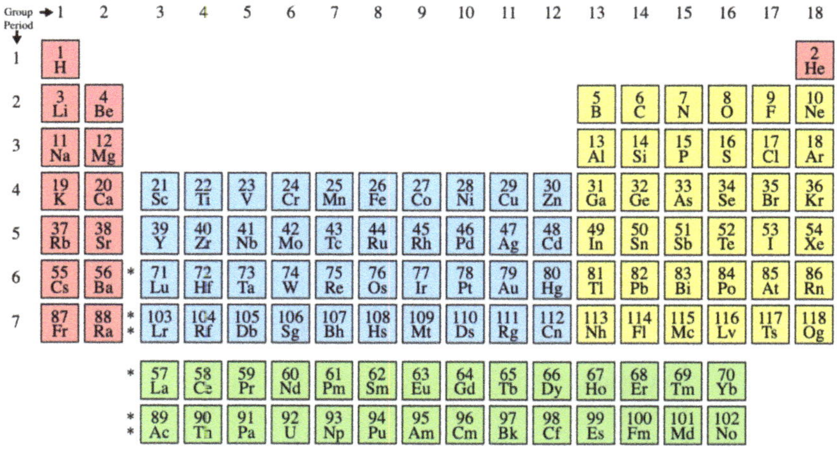

Figure 1.19. The periodic table and organization of the 118 known chemical elements; 90 found in Nature, 28 produced by nuclear reactions in the laboratory. The element 'boxes' are color coded by the subshell filling for the electrons: red—s ($l = 0$), yellow—p ($l = 1$), blue—d ($l = 2$), green—f ($l = 3$). Should synthesis of even heavier elements be achieved, the 'super-actinides' may be discovered: these elements will involve electrons filling the g ($l = 4$) subshell. This periodic table image has been obtained by the authors from the Wikimedia website where it was made available under a CC BY-SA 4.0 licence. It is included within this chapter on that basis. It is attributed to Double sharp.

Shell structure for atoms is also evident in atomic ionization potentials (electron 'separation' energies), as shown in figure 1.20. Indeed, shell structure in nuclei and shell structure in atoms result from the same underlying quantum mechanics: the many bodies (nucleons in nuclei, electrons in atoms) behave as independent particles confined by an *average single-particle potential.* While nuclei also exhibit correlations, which manifestly are not independent-particle behavior, independent-particle behavior was the fundamental insight into nuclear structure that launched the quantum mechanical description of nuclei. Basic details are given in chapter 2.

The study of nuclei has been a branch of science ever since the observations of Hans Geiger and Ernest Marsden with respect to scattering of alpha particles [19, 20] and the interpretation [21] of these observations by Ernest Rutherford: that atoms have nearly all of their mass (99.97%) concentrated in a volume of space that is only 10^{-15} of the atomic volume. With the discovery that nuclei contain neutrons [22] by James Chadwick and the variety of nuclear processes—radioactive decay and reactions—there has been a steady unfolding saga under the title of nuclear physics. It is important to know, as nuclear physicists (and nuclear chemists), what the key steps were. We recommend to the Reader an excellent essay, written by Bethe [23] at the turn of the millennium, which provides a concise perspective on the science of the nucleus as it unfolded in the 20th century.

1.1 Exercises

The relative challenge and time investment associated with the exercises is indicated from * to ***. Exercises marked with '**E**' require interactive access to figures or

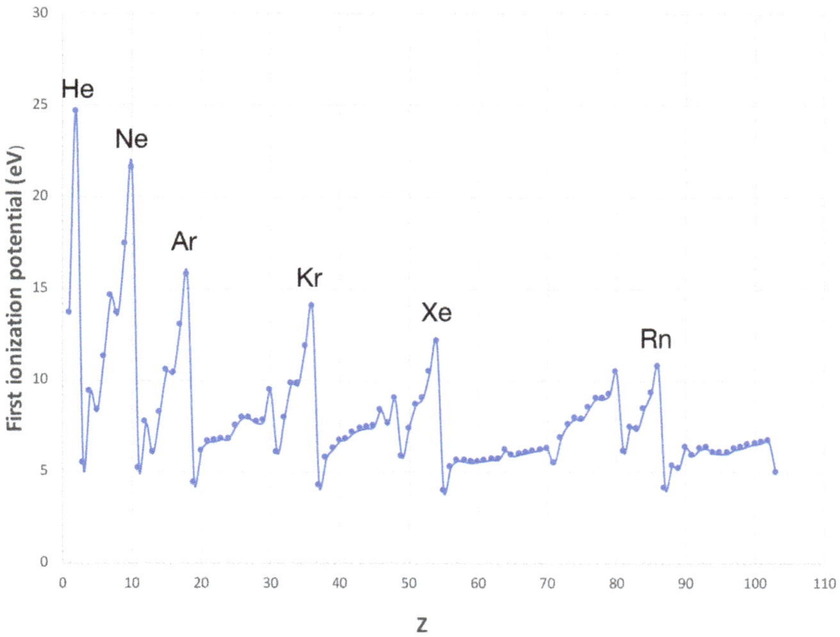

Figure 1.20. First ionization potentials for atoms, i.e. the energy needed to remove a single electron from the neutral atom. Energies are given in electronvolts (eV).

watching included video exercises and will only have such functionality in the e-book version, available at https://iopscience.iop.org/book/978-0-7503-2674-2.

1-1. In equation (1.3), R_0 is conventionally taken to be 1.2 fm. This is a rounded-off value for R_A defined in figure 1.2, i.e. it is the effective radius of the nucleus if viewed as a liquid drop with a constant density and a sharp surface. Calculate $R_0A^{1/3}$ for the nuclei depicted in figure 1.3. Note: from equation (1.3), these values are for $1.29\langle r^2\rangle^{1/2}$ and so they will exceed the 'half density' values for the data presented. **

1-2. By integrating over the volume of a nucleus of constant density and a sharp surface of radius R_A, obtain the factor $(5/3)^{1/2}$ in equation (1.3). (Hint: the quantity is the average of r^2 over the volume of a sphere.) **

1-3. Using equation (1.3), obtain a relationship for $\delta\langle r^2\rangle$ suitable for comparison with the isotope-shift data shown in figure 1.4 (ignore the trend shown for the heaviest yttrium isotopes depicted). **

1-4. Using the data in table 1.1:
   (a) How much gain in binding energy would be needed for ^{5}He to be bound in its ground state with respect to disintegration into ^{4}He + n? *
   (b) How much gain in binding energy would be needed for ^{8}Be to be bound in its ground state with respect to disintegration into ^{4}He + ^{4}He? *
   (c) How much energy is needed to separate a neutron from ^{6}He? *

1-5. It was predicted by Fred Hoyle that, since ^{8}Be is unbound, three ^{4}He nuclei must fuse to form ^{12}C in stellar nucleosynthesis and that there must be an excited state in the ^{12}C nucleus just above the energy for ^{12}C to disintegrate into three ^{4}He nuclei. What is the approximate energy of this state? (Against great skepticism, his prediction proved correct and, with his recognition for this and other astrophysical research, he was recognized by being dubbed Sir Fred on January 1st, 1972. The state in ^{12}C is the only quantum state in the entirety of nuclear physics named after a person; we call it the Hoyle state.) **

1-6. In the Sun, hydrogen fuses into ^{4}He. The age of the Sun is 4.6×10^{9} y and its mass is 1.99×10^{30} kg. The energy output of the Sun is 4×10^{20} MW.
   (a) Using an electron mass of 0.51 MeV c^{-2}, how much energy is produced per atom of helium produced? *
   (b) How many kilograms of hydrogen are being fused into helium per second? *
   (c) What percentage of its mass has been fused into helium so far? *
   (d) If the energy source of the Sun were chemical, how long would the Sun shine for the given energy output? Assume an energy yield of 10 eV per hydrogen atom. *

1-7. Uranium can undergo fission, e.g. $^{238}\mathrm{U} \rightarrow {}^{138}\mathrm{Te} + {}^{100}\mathrm{Zr}$. How much energy is released per kilogram? **

1-8. Which beta decay processes can occur for the mass 96 nuclei given in table 1.1? (The isotope ^{96}Zr occurs in Nature with a natural abundance of 2.8%. Take a look at ENSDSF if you find this puzzling.)

1-9. The free neutron undergoes β^{-} decay to hydrogen. How much energy is released per neutron? *

1-10. Why doesn't ^{2}H decay into two protons (plus an electron)? *

1-11. For an atom of ^{2}H, what is the ratio of electronic binding energy to nuclear binding energy? *

1-12. Using equation (1.3) with $R_0 = 1.2$ fm and Coulomb's law, for the fission fragments ^{100}Zr and ^{138}Te, if their surfaces are just touching, calculate the height of the barrier in figure 1.18. Depict your answer in a manner similar to this figure, i.e. include your answer to exercise 1-7. **

1-13. The Bethe–Weizsäcker formula for the binding energy of nuclei is expressed as

$$B(A, Z, N) = a_1 A + a_2 A^{2/3} + a_3 \frac{Z^2}{A^{2/3}} + a_4 \frac{(N - Z)^2}{A}, \tag{1.6}$$

where (approximately)

$$a_1 = 15.68 \text{ MeV}, \ a_2 = -18.56 \text{ MeV}, \ a_3 = -0.717 \text{ MeV}, \ a_4 = -28.1 \text{ MeV}, \tag{1.7}$$

and the four terms are called the 'volume', 'surface', 'Coulomb', and 'symmetry' terms, respectively. Note that there is no pairing term in equation (1.7). Using equation (1.2), what would be the value of Z for the most stable isotope with:

(a) A = 133? Compare with figure 1.16.
(b) A = 134? Compare with figure 1.17.
(c) A = 294?
(d) A = 400?

[Hint: maximize B versus Z at constant A; and substitute $N = A - Z$ before starting!] ***

1-14. Explore the contribution of the four terms in the Bethe–Weizsäcker binding energy formula presented in exercise 1-13 by setting, in turn, the parameters a_1 to a_4 to zero. How would such changes compress or expand the reach of the table of isotopes? Explore particularly the effects on the stability of the actinide isotopes and the relative stability of even–even, odd–A and odd–odd isotopes. ***

1-15. For two nuclei with $(A, Z) = (A_1, Z_1)$ and (A_2, Z_2), if they have uniform density and sharp surfaces, and they are just in contact, obtain a formula for the Coulomb repulsion energy between them in MeV. Use $R = R_0A^{1/3}$ with $R_0 = 1.2$ fm. Assume that their charges act as if concentrated entirely at their centres of mass. Express your answer in the form

$$E_{\text{Coulomb}} = \frac{VZ_1Z_2}{(A_1^{1/3} + A_2^{1/3})}, \tag{1.8}$$

i.e. find V, expressed in MeV. **

1-16. How many elements will there be in the first super-actinide series? (See the definition of the super-actinide series in the caption to figure 1.19.) **

1-17. Naively (ignoring relativistic effects in atomic structure, which become very large as Z increases), what will be the atomic number of the next inert gas? **

1-18. Estimate the most stable mass with respect to beta decay of your answer to exercise 1-17. **

1-19. Using data from the AME, estimate the location of the neutron drip line for odd-mass Sn isotopes. **

1-20. What fraction of the nuclear volume is 'empty' space? (Use a proton radius of 0.85 fm and assume the same radius for a neutron.) **

1-21. In figure 1.5, the hyperfine splitting is presented for the $J = 3/2$ and $J = 5/2$ electron orbitals in ^{178m}Hf. The hyperfine quantum number, $\mathbf{F}$, is defined by

$$\mathbf{F} = \mathbf{I} + \mathbf{J}, \tag{1.9}$$

where $\mathbf{J}$ is the electron total angular momentum and $\mathbf{I}$ is the nuclear spin. Using this vectorial coupling, deduce the possible values of F for each of the electron hyperfine substates and verify that these exist in figure 1.5. Verify with reference to figure 1.5 that the selection rule for hyperfine transitions is $\Delta F = 0, 1$. **

1-22. Figure 1.11 presents S_{2n}—the two-neutron separation energy for the Ca isotopic chain. The two-neutron separation energy can be conveniently expressed in terms of mass excess, i.e. the difference between the actual mass and the mass number in atomic mass units, as

$$S_{2n} = \Delta M(Z, N-2) + 2\Delta M_n - \Delta M(Z, N). \tag{1.10}$$

Use the mass excess data found in the AME to make your own plot of S_{2n} as a function of A for the ${}_{28}$Ni isotopes from ^{52}Ni to ^{80}Ni. (Note that for many of the more exotic isotopes what is presented in the AME is an estimate not a measured value.) Identify in your plot the location of the shell closures. **

1-23. For those reading the e-book, explore the video-based exercise in figure 1.21, which relates to predicting the location of the neutron drip-line and encourages engagement with the data on nuclear masses found in

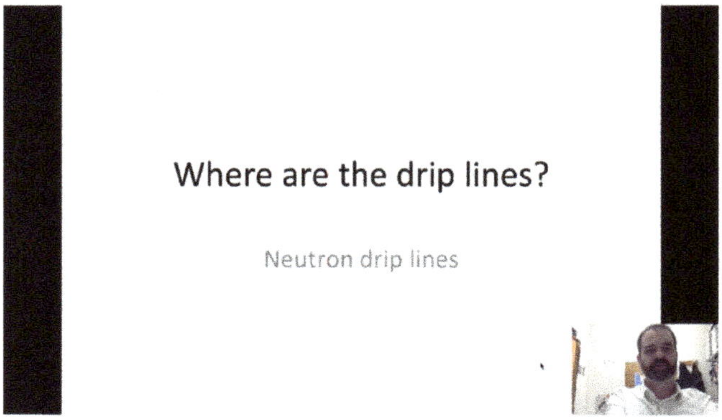

Figure 1.21. Video exercise: Predicting the location of the neutron drip line. Video available at https://iopscience.iop.org/book/978-0-7503-2674-2.

Figure 1.22. Video exercise: Predicting the location of the proton drip line. Video available at https://iopscience.iop.org/book/978-0-7503-2674-2.

the AME. The powerpoint presented in the video can also be accessed here: [https://iopscience.iop.org/book/978-0-7503-2674-2]. ** **E**

1-24. For those reading the e-book, explore the video-based exercise in figure 1.22, which relates to predicting the location of the proton drip-line and encourages engagement with the data on nuclear masses found in the AME. The powerpoint presented in the video can also be accessed here: [https://iopscience.iop.org/book/978-0-7503-2674-2]. ** **E**

References

[1] Rowe D J and Wood J L 2010 *Fundamentals of Nuclear Models: Foundational Models* (Singapore: World Scientific)

[2] Hahn B, Ravenhall D G and Hofstadter R 1956 High-energy electron scattering and the charge distributions of selected nuclei *Phys. Rev.* **101** 1131

[3] Yennie D R, Ravenhall D G and Wilson R N 1954 Phase-shift calculation of high-energy electron scattering *Phys. Rev.* **95** 500

[4] Herman R C and Hofstadter R 1960 *High-Energy Electron Scattering tables* (Palo Alto, CA: Stanford University Press)

[5] Frois B and Papanicolas C N 1987 Electron scattering and nuclear structure *Ann. Rev. Nucl. Part. Sci.* **37** 133

[6] Cheal B *et al.* 2007 The shape transitions in the neutron-rich yttrium isotopes and isomers *Phys. Lett.* B **645** 133

[7] Cheal B and Flanagan K T 2010 Progress in laser spectroscopy at radioactive ion beam facilities *J. Phys.* G **37** 113101

[8] Wang M *et al.* 2017 The AME2016 atomic mass evaluation (II). tables, graphs and references *Chin. Phys.* C **41** 030003 (Note added in proof: following the completion of this book, AME2020 was published in 2021, Chin. Phys. C 45 030003)

[9] Angeli I and Marinova K P 2013 table of experimental nuclear ground state charge radii: An update *At. Data Nucl. Data tables* **99** 69

[10] Goeppert-Mayer M 1949 On Closed shells in nuclei. II *Phys. Rev.* **75** 1969

[11] Haxel O, Hans J, Jensens D and Suess H E 1949 On the 'magic numbers' in nuclear structure *Phys. Rev.* **75** 1766

[12] Schüler H and Schmidt T 1935 Über Abweichungen des Atomkerns von der Kugelsymmetrie *Z. Phys.* **94** 457

[13] Casimir H B G 1936 On the interaction between atomic nuclei and electrons *Teyler's Tweede Genootschap* **11**

[14] Heyde K and Wood J L 2016 Nuclear shapes: from earliest ideas to multiple shape coexisting structures *Phys. Scr.* **91** 083008

[15] Lieb K P 2001 Theodor Schmidt and Hans Kopfermann—pioneers in hyperfine physics *Hyp. Interact* **136** 783

[16] Stone N J 2016 Table of nuclear electric quadrupole moments *At. Data Nucl. Data tables* **111-112** 1

[17] Ahn D S *et al.* 2019 Location of the neutron dripline at fluorine and neon *Phys. Rev. Lett.* **123** 212501

[18] Oganessian Y T and Utyonkov V K 2015 Super-heavy element research *Rep. Prog. Phys.* **78** 036301

[19] Geiger H and Marsden E 1909 On a diffuse reflection of the alpha-particles *Proc. R. Soc.* A **82** 495
[20] Geiger H 1910 The scattering of alpha-particles by matter *Proc. R. Soc. Lond.* A **83** 492
[21] Rutherford E 1911 The scattering of alpha and beta particles by matter and the structure of the atom *Philos. Mag. (Abingdon)* **21** 669
[22] Chadwick E 1932 The existence of a neutron *Proc. R. Soc. Lond.* A **136** 692
[23] Bethe H A 1999 Nuclear physics *Rev. Mod. Phys.* **71** S6

Chapter 2

Nuclear excitation patterns

Excited states in nuclei and their quantum number labels are introduced. Bound and unbound states and lifetimes are mentioned. Basic techniques for determination of excitation energy, spin and parity are sketched; this includes gamma–gamma angular correlations and internal-conversion electron spectroscopy. The philosophy of approach to the organization of nuclear data is laid down. Signatures of behavior are described, especially rotation. The basic patterns of energy and electromagnetic decay, including Weisskopf estimates of strength, are introduced; electric quadrupole transitions are emphasized. Mention is made of Coulomb excitation, multipole mixing ratios, and electric monopole transitions. The concept of isospin is illustrated. Attention is drawn to accessible sources of nuclear data.

Concepts: direct and indirect measurement of excitation energies, gamma–gamma angular correlations, internal-conversion electrons, level density and complexity, $E(2_1^+)$, $R_4 = E(4_1^+)/E(2_1^+)$, rotational energy pattern, closed shell energy gap, electric quadrupole transition strength—$B(E2)$, Weisskopf estimates, isospin symmetry.

Each nucleus exhibits a characteristic pattern of excitation. Excited states are labeled almost always by at least their energy, usually by a spin and parity, sometimes by a magnetic moment, less often by a quadrupole moment, and occasionally by a mean-square charge radius. All of these quantities are quantum mechanical observables. Often, excited nuclear states are labeled by a half-life (or a lifetime). Other properties are introduced as a more detailed focus is made on nuclear properties.

Excitation energies can be determined via differences in energies of particles from a monoenergetic beam that populates the excited state (and the ground state) in a reaction, as shown in figure 2.1(a), or by measurement of the energy of radiation emitted when the excited state decays (following some excitation process), as shown in figure 2.2(a)–(d). Mass-energy is conserved. Some nuclei only have a finite number of bound states. A few nuclei have only a single bound state, e.g. ^{2}H. Whether or not an excited state is 'bound' is an issue of decay with respect to emission of nucleons or

doi:10.1088/978-0-7503-2674-2ch2

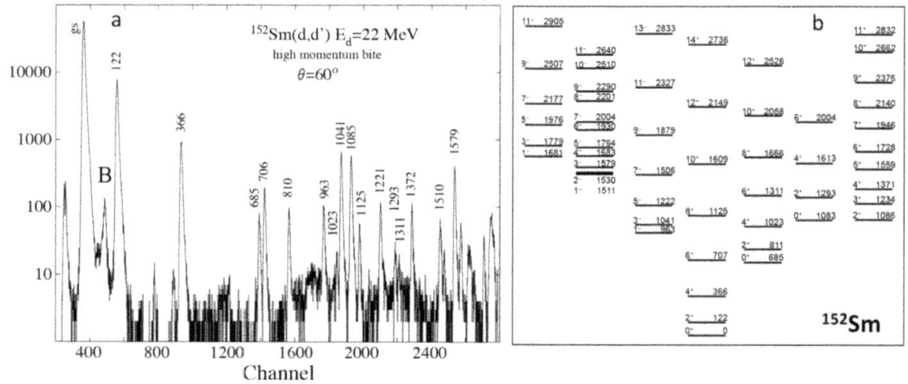

Figure 2.1. (a) and (b) An example of the direct observation of excited states in a nucleus by measurement of energy differences for inelastically scattered particles from a monoenergetic beam incident on a target containing the nucleus of interest. (a) The data are for the nucleus ^{152}Sm and the scattering of deuterons. *B* indicates a background line due to a target impurity. (b) The lowest excited states in ^{152}Sm arranged in rotational bands. The process favors low-spin states of natural parity (see text). The ground state (energy zero) is labeled as 'gs'. The deuteron scattering data are presented courtesy of P E Garrett. Video-based exercises relating to how to interpret and order the rotational structures in ^{152}Sm are found in exercise 2-11.

nucleon clusters and the lifetime of the state. If the lifetime (τ) is very short, the level is specified by a decay width (ΔE) given by

$$\Delta E\tau \sim \hbar. \tag{2.1}$$

For details of equation (2.1), especially the derivation of the relationship, see [1] sections 9.4 and 9.9.

Ground-state spins can be determined by Stern–Gerlach experiments whereby an atomic beam passes through an inhomogeneous magnetic field and the magnetic substates, $M_J = +J, +J - 1, +J - 2, \ldots, -J$, i.e. $2J + 1$ values are dispersed into distinct groups of events recorded, e.g. on a photographic plate. Counting the $2J + 1$ groups of events is the most direct indication of a spin value, J. Spin is fundamentally a conserved quantity because nuclei are finite many-body systems: no matter how complex the coupling of spins of constituent particles and their angular momenta, their total resultant spin is a good quantum number. Nuclei can be regarded as conservative systems, i.e. their properties are not influenced in any way by their environment. The nucleus, as the constituent of an atom, can be studied by atomic hyperfine spectroscopy. From knowledge of the electronic contribution to the total atomic spin, this yields the total nuclear spin, cf. figure 1.5.

Assigning parity to a nuclear energy eigenstate is a more subtle process. Parity is either 'even' (+) or 'odd' (−), indicating the symmetry of the total nuclear state vector with respect to spatial inversion. We speak of positive- and negative-parity states. This is also an observable for a finite many-body quantum system. Most commonly at low excitation energy in even–even nuclei (all spin values are integers) even-spin states have positive parity and odd-spin states have negative parity, i.e. parity, $\pi = (-1)^J$ (called a state of natural parity). We will mainly encounter

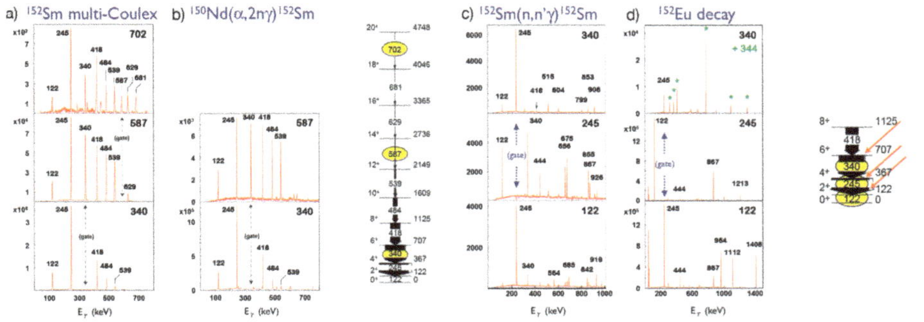

Figure 2.2. (a)–(d) Examples of the indirect observation of excited states in a nucleus by measurement of γ-ray transition energies. The four data sets are for the nucleus ^{152}Sm with population of excited states by four different processes, all using measurement of coincidences between pairs of de-exciting γ-rays: (a) multi-step Coulomb excitation of a beam of ^{152}Sm ions incident on a ^{208}Pb target with coincidence 'gates' on the 340 keV, $6^+_1 \to 4^+_1$, 587 keV, $14^+_1 \to 12^+_1$, and 702 keV, $20^+_1 \to 18^+_1$ transitions. (b) Production following fusion of α particles (^{4}He nuclei) in an energetic beam incident on ^{150}Nd target nuclei with 'evaporation' of two neutrons. The evaporation is statistical, with other evaporation 'channels' (evaporation of one and three neutrons, or one proton); but γ-ray coincidences are highly selective, with gates shown as per (a). (c) Inelastic scattering of neutrons from a ^{152}Sm target with coincidence gates on the 122 keV, $2^+_1 \to 0^+_1$, 245 keV, $4^+_1 \to 2^+_1$, and 340 keV, $6^+_1 \to 4^+_1$ transitions. (d) Population of excited states via β^+ and electron capture decay of ^{152}Eu with coincidence gates set as per (c). Note the very similar pattern of γ-ray intensities in part (a) and (b) for the 587 keV gate—all of the γ-ray lines at lower energies inevitably follow the transition in the gate. The attenuation of the 122 keV γ-ray intensity is due to a combination of detector efficiency decreasing at low energy and to the internal conversion process (see figure 2.4); $\alpha_{TOT}(122\ \text{keV}, Z = 62) = 1.16$. The population of states in (a) and (b) shows the preference for high-spin population due to the high probability that a beam particle does not impact a target particle with zero transfer of angular momentum. The population of states in (c) and (d) shows a preference for low-spin population; (c) because the neutron-beam energy is low and (d) because the spin of the β-decaying nucleus, ^{152}Eu is $J = 3$ (and β decay is dominated by spin selectivity with $\Delta J = 0, 1$). The ground-state band is shown for orientation; also see figure 2.1, consultation of which enables a number of γ-ray transitions to be located. Note that the 340 keV gate in the ^{152}Eu decay study contains 344 keV γ-rays from the β^- decay of this nucleus to ^{152}Gd $[E(2^+_1) = 344\ \text{keV}]$. This shows that one can obtain a very detailed view of possible excited states in a given nucleus by making use of a number of experiments with probes that are sensitive to properties of the atomic nucleus thereby acting in a complementary way. The spectra were made from a variety of data sets by W D Kulp and are used with permission. Video-based exercises relating to how to interpret and order the rotational structures in ^{152}Sm are found in exercise 2-11.

natural-parity states, and will pay particular attention to low-energy unnatural-parity states. But in totality, there are as many odd-parity states as even-parity states in excitations across all nuclei.

Spins of excited states can be determined, e.g. by angular distributions of scattered particles from a beam relative to the beam direction, by angular distributions of radiations following population of the state by a reaction (with a beam of particles) or from a polarized source. Changes in spin can be determined by angular correlations between sequentially emitted radiations, most notably between pairs of emitted γ-rays, as illustrated in figure 2.3. A more subtle but common source of information on spin changes is the so-called internal conversion process: the relative decay probability mediated by atomic electrons versus γ-ray emission is measured.

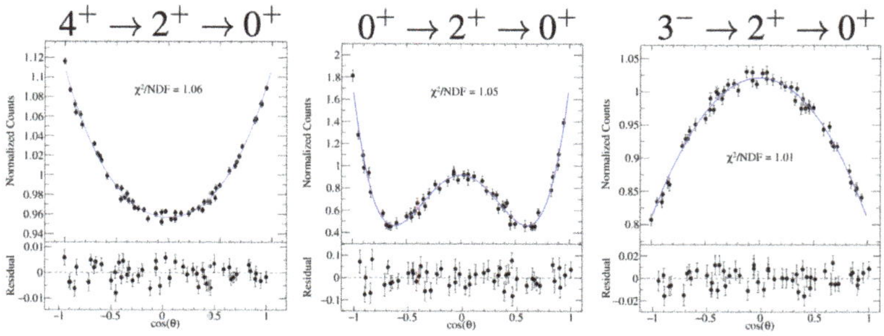

Figure 2.3. Angular correlations between pairs of γ-rays, observed in coincidence as a function of the cosine of the angle between detectors, for the decay spin sequences: (left) $4^+ \rightarrow 2^+ \rightarrow 0^+$; (middle) $0^+ \rightarrow 2^+ \rightarrow 0^+$; (right) $3^- \rightarrow 2^+ \rightarrow 0^+$. (Reprinted from [2], copyright (2019), with the permission of Elsevier.)

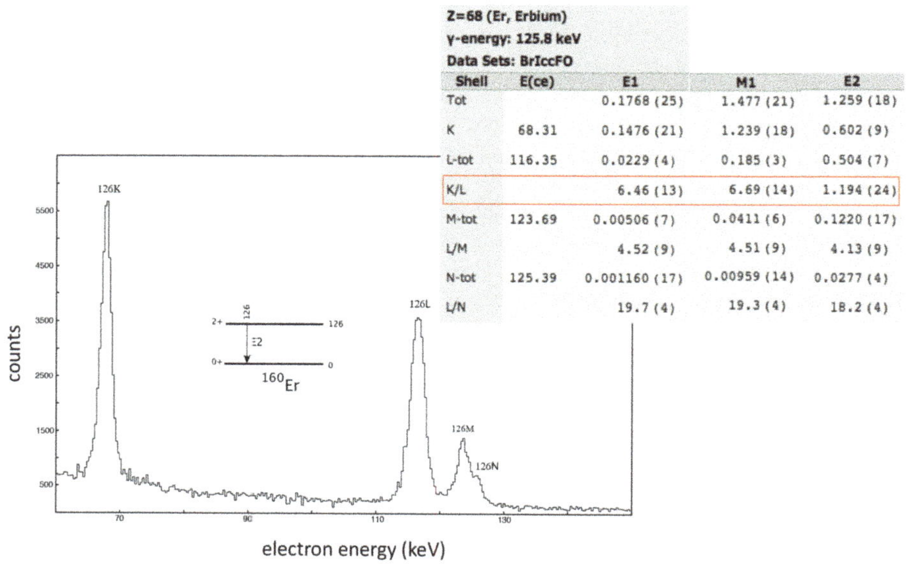

Figure 2.4. An electron spectrum observed in the decay ^{160}Tm $\rightarrow$ ^{160}Er, with the electron lines corresponding to internal conversion decay of the 2_1^+ state in ^{160}Er. The multiplet of electron lines results from the various probabilities of the conversion process occurring in the *K*, *L*, *M*, and *N* atomic electron shells. The electron energies reflect the different atomic electron binding energies where $E_{ce} = E_\gamma - B_i$, $i = K, L, M, N$. Note that the electron detector resolution is not high enough to resolve the subshell contributions—L_1, L_2, L_3; M_1, M_2, etc. The inset shows the comparison of the observed subshell intensities with theory (see text). The *K*/*L* ratio indicates that the multipolarity of the transition is *E*2. Note that the *L* line is 'wide', because the *L* subshells have different energies, and so the height of the *L* line is less than an expected 4500 counts as per the theoretical value for *E*2; but the areas confirm the assignment. The figure is from an unpublished data set of one of the authors (JLW).

The resulting internal conversion coefficients reflect well-defined probabilities for K, L_1, L_2, L_3, M_1, ... shell conversion as a function of atomic number (Z), transition energy, spin and parity change (mulitpolarity) for the specific nuclear transition, as illustrated in figure 2.4. Such data provide a powerful basis for constructing a nuclear level scheme with spin-parity assignments for excited states. However, it should be noted that spin changes are changes in a vector quantity and the combination of two spin vectors $\mathbf{J}_1$ and $\mathbf{J}_2$ can in general result in any value $|\mathbf{J}_1 - \mathbf{J}_2|$, $|\mathbf{J}_1 - \mathbf{J}_2| + 1$, $|\mathbf{J}_1 - \mathbf{J}_2| + 2$, $\ldots|\mathbf{J}_1 + \mathbf{J}_2|$. Selectivity from this range of possibilities can occur under experimental circumstances that are discussed later.

A number of universal features of states in nuclei are evident:

1. All even–even nuclei have ground-state spins $J = 0$.
2. Most even–even nuclei have first excited states with spin $J = 2$.
3. All even-mass nuclei have integer spin values for all states and all odd-mass nuclei have odd-half integer spin values for all states.
4. The low-energy level density is: odd–odd nuclei $\gg$ odd-Z $\sim$ odd-N nuclei $>$ even–even nuclei. This is illustrated in figure 2.5.

Figure 2.5 provides a major caution on how to proceed to an understanding of excited states in nuclei. One should start with even–even nuclei because the lowest energy states have lower density than for odd nuclei and much lower density than for odd–odd nuclei, and therefore basic patterns are easier to discern. There is no established procedure for this task. Another popular approach is to start at doubly closed shells and progressively 'add nucleons'. We will arrive at such nuclei as a late step in the approach that we adopt herein. The primary motivation is that nuclei at and near doubly closed shells are very few in number and, it is emerging, their lowest excited states can exhibit subtle features that need a global perspective for understanding. Most notably, we reiterate, the role of quantum mechanical correlations is universally present in nuclei.

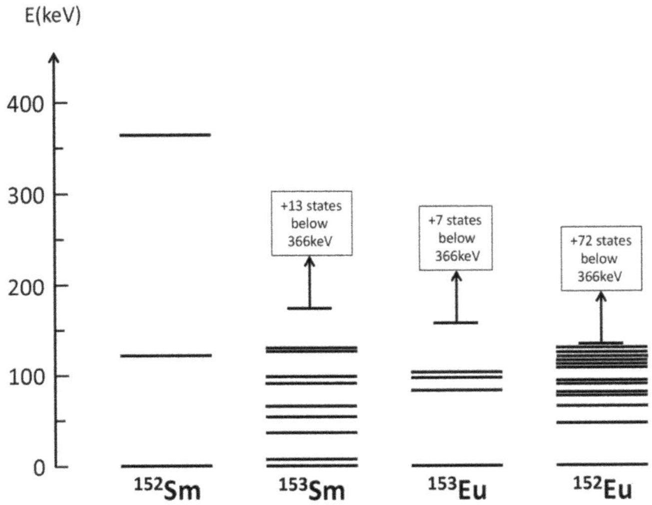

Figure 2.5. Comparison of the number of excited states, observed at low energy, for the neighboring even–even, odd-Z, odd-N, and odd–odd nuclei, ^{152} Sm, ^{153}Eu, ^{153} Sm, and ^{152}Eu. The density of low-energy excited states critically impacts the difficulty of conducting nuclear spectroscopy: in practical terms this is a decision-making process in the face of escalating complexity and real-time data reduction tasks.

2.1 Basic features of excited states in even–even nuclei

Even–even nuclei are chosen for an introduction to excited states in nuclei because they exhibit recognizable patterns that recur across the entire mass surface. Thus, their excitations are used as a basic guide to nuclear structure.

2.1.1 Excitation energies

The single most pervasive feature of nuclear excitations is the occurrence of first excited states in even–even nuclei with spin-parity $J^{\pi} = 2^{+}$. There are only a few exceptions to this observation and they will be an important issue of focus later. A 'helicopter' view of first excited 2^{+} state energies across the nuclear chart is shown in figure 2.6 along with 'maps' for two broad mass regions in figure 2.7 and 2.8.

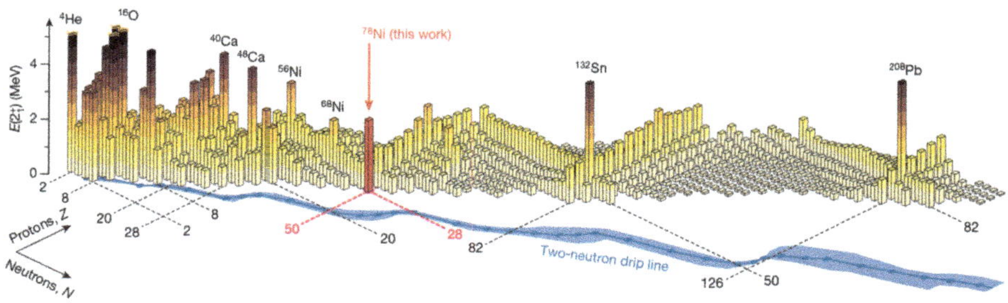

Figure 2.6. A helicopter view of the 2_1^+ excitation energies across the nuclear chart. Note that for some light nuclei such as ^{4}He and ^{16}O, the 2_1^+ energy lies significantly off the z-axis scale used. Reprinted by permission from Springer Nature: [3], copyright (2019).

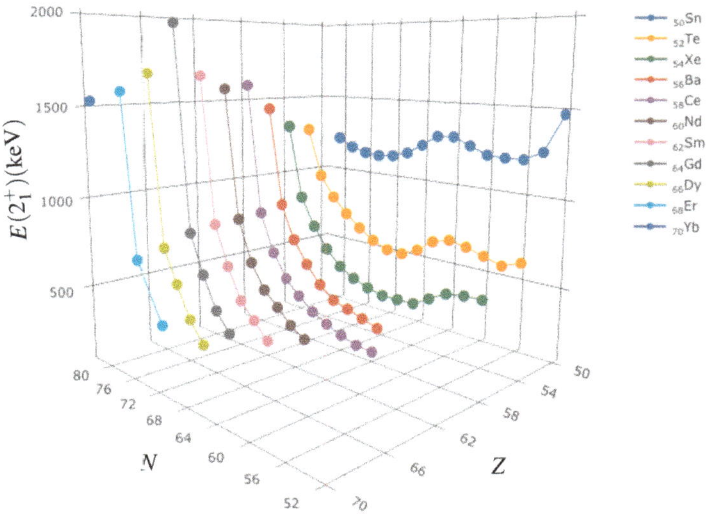

Figure 2.7. Map of E(2_1^+) values in keV for nuclei bounded by $Z \geqslant 50$ and $N \leqslant 82$. The high 'ridges' for $Z = 50$ (tin) isotopes and $N = 82$ isotones are characteristic of nuclei with closed shells. Note that the double-closed shell nucleus ^{132}Sn has a value of 4041 keV, which is not shown as it is off the scale used. This is a 'southeastwards view' of the Chart of the Nuclides. Animation available at https://iopscience.iop.org/book/978-0-7503-2674-2.

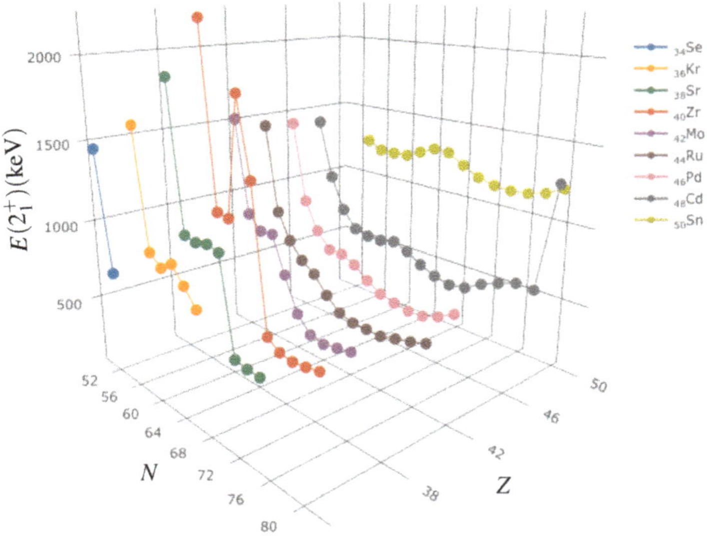

Figure 2.8. Map of E(2_1^+) values in keV for nuclei bounded by $N \geqslant 50$ and $Z \leqslant 50$. Note that the double-closed shell nucleus ^{132}Sn has a value of 4041keV, which is not shown as it is off the scale used. This is a 'northwestwards view' of the Chart of the Nuclides. Animation available at https://iopscience.iop.org/book/978-0-7503-2674-2.

One notes that the variation in energy is smooth with high $E(2_1^+)$ values at $N = 50$, $Z = 50$ and $N = 82$. These are closed shell numbers.

Regions where $E(2_1^+)$ values are lowest are far removed from the closed shell numbers. The lowest known values occur in the actinide region around ^{250}Cf ($Z = 98$, $N = 152$, 42.72 keV) and in the rare earth region around ^{170}Dy ($Z = 66$, $N = 104$, 71.47 keV). Without exception, nuclei with the lowest energy 2_1^+ states possess second excited states with $J^\pi = 4^+$. The energy ratio, $R_4 := E(4_1^+)/E(2_1^+)$ in all these nuclei is close to 3.33. This is characteristic of an axially symmetric rotor with a plane of reflection symmetry perpendicular to the axis of rotational symmetry. The R_4 values for the rare earth region are shown in figure 2.9, and for two other mass regions in figures 2.10 and 2.11. Indeed, the excitation spectrum for an axially symmetric rotor model possesses the states, expressed as ratios to $E(2_1^+)$

$$J^\pi(E_J/E_2) = 4^+(3.333),\ 6^+(7.000),\ 8^+(12.000),\ \dots, \tag{2.2}$$

and nuclei in the rare earth and actinide regions with the lowest $E(2_1^+)$ values exhibit excited states with energies very close to the sequence given in equation (2.2). These energies follow from the simple formula

$$E = \frac{\hbar^2 I(I+1)}{2\mathcal{J}}, \tag{2.3}$$

where $\mathcal{J}$ is a moment of inertia parameter. (Note that the symbols 'I' and 'J' are used interchangeably to represent spin and angular momentum, and I is usually the preferred symbol where rotations are involved.) A useful plot is for

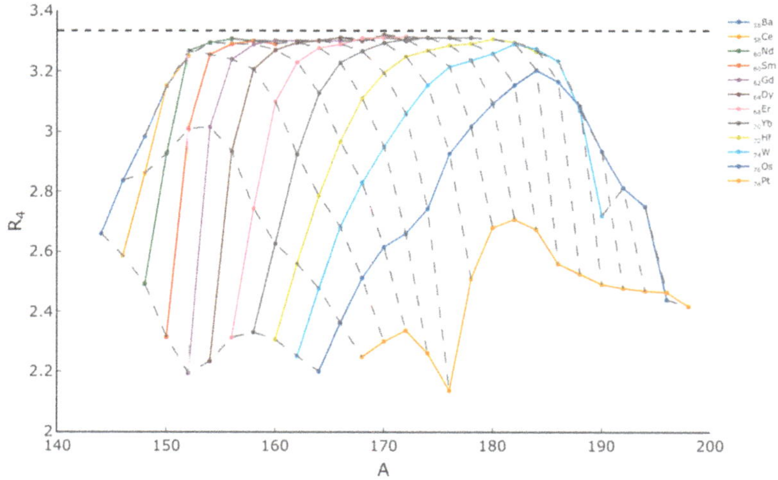

Figure 2.9. Map of $R_4 := E(4_1^+)/E(2_1^+)$ values for nuclei in the rare earth region. The close approach of many of these nuclei to $R_4 = 3.333$ (dashed horizontal line) is interpreted as due to deformation, as emerges from the simple rotor model of deformed nuclei, discussed in chapter 3. Solid lines connect isotopic chains and dashed lines connect isotonic chains. Animation available at https://iopscience.iop.org/book/978-0-7503-2674-2.

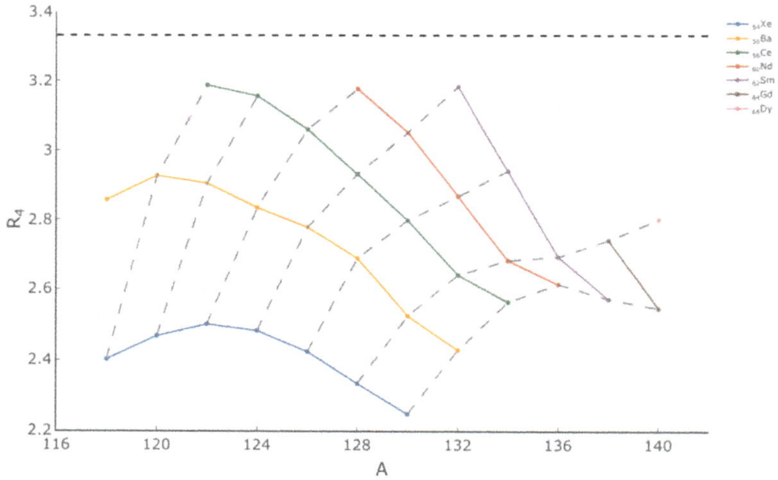

Figure 2.10. Map of R_4 values for nuclei in the open-shell region $Z \geqslant 54$, $N \leqslant 74$, $118 \leqslant A \leqslant 140$. The mid-shell point is located in the unexplored region of ^{132}Dy, cf. figure 2.7. For other details, see figure 2.9. Animation available at https://iopscience.iop.org/book/978-0-7503-2674-2.

$$r_6 := \frac{E(6_1^+) - E(4_1^+)}{E(2_1^+)} \text{ versus } r_4 := \frac{E(4_1^+) - E(2_1^+)}{E(2_1^+)}, \tag{2.4}$$

and the nuclei with the largest R_4 values are shown in figure 2.12. Note $r_4 = R_4 - 1$ and $r_6 = R_6 - R_4$. The use of r_6 versus r_4 plots instead of R_6 versus R_4 plots is preferred because the measured quantities are γ-ray transition energies, i.e. differences

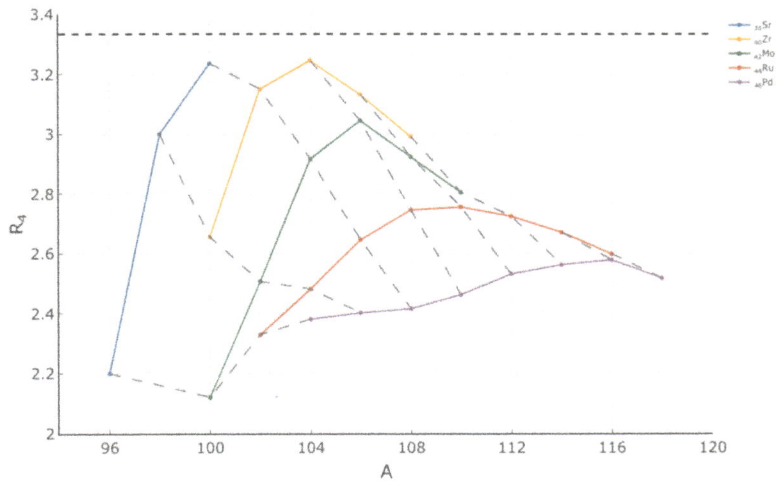

Figure 2.11. Map of R_4 values for nuclei in the open-shell region $Z \geqslant 38$, $N \leqslant 46$. This region is centred on the mid-shell shown in figure 2.8. For other details, see the caption to figure 2.9. Animation available at https://iopscience.iop.org/book/978-0-7503-2674-2.

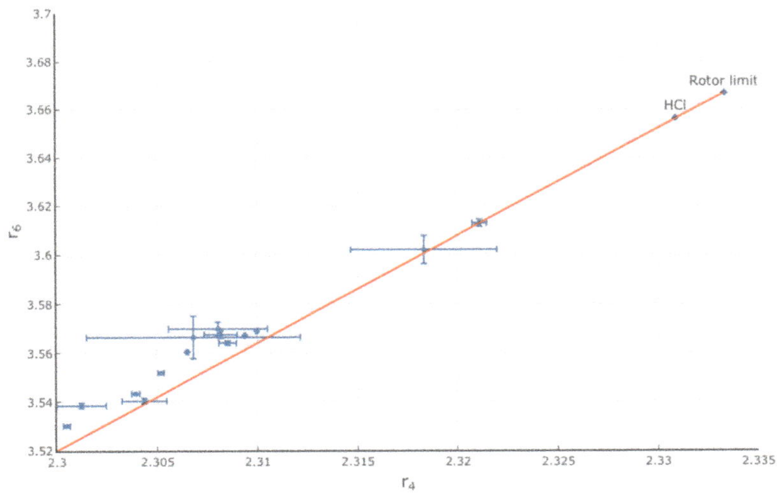

Figure 2.12. Plot of $[E(6_1^+)–E(4_1^+)]/E(2_1^+) := r_6$ versus $[E(4_1^+)–E(2_1^+)]/E(2_1^+) := r_4$. The simple rotor limits are at $r_4 = 2.333\,3$, $r_6 = 3.666\,7$. For comparison, the rotations of the diatomic molecule, HCl are shown (see figure 5.17 and exercise 5-6). The error bars on some of the data points reflect the uncertainties in the γ-ray transition energies used to compute the ratios; if the datum is a point, the uncertainties are too small to depict at the scale used. The 'most rotational' ground-state band known occurs in ^{250}Cf. The red line corresponds to the relationship $E = AI(I + 1) + BI^2(I + 1)^2$, see exercise 2-1. Animation available at https://iopscience.iop.org/book/978-0-7503-2674-2.

in the level energies, and so the uncertainties in the plotted quantities are easier to depict. Some details of the axially symmetric rotor are given shortly.

Regions where $E(2_1^+)$ values are highest are for closed shell nuclei, with extremely high values for doubly closed shell nuclei. The excitations in the even-mass Sn

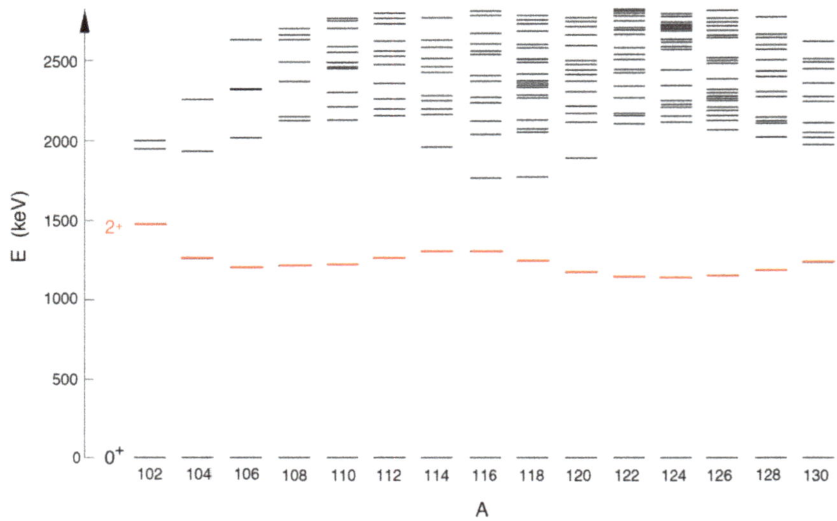

Figure 2.13. Illustration of the basic excitation pattern observed in closed shell nuclei—example of the $Z = 50$ (tin) isotopes. Closed shell nuclei are generally characterized by a significant energy gap, with only one or very few excited states at low energy. However, starting at or just above the second excited state, a very high-level density begins: this leads to the need to handle spectroscopic complexity. The gap is due to pairing correlations and this is discussed in detail in chapter 4.

isotopes ($Z = 50$) are shown in figure 2.13. The first excited state has $J^{\pi} = 2^+$ and occurs at a fairly constant energy of 1.2 MeV, but then there is the onset of a dense spectrum of excited states above about 2 MeV. Note that across the entire neutron shell ($50 < N < 82$) the excitations appear approximately constant, independent of the neutron number. There is not a minimum in the $E(2_1^+)$ values at mid-shell, $N = 66$. In due course, we will identify excitation patterns in singly closed shell nuclei.

2.1.2 Electromagnetic decay properties

The systematic decay strengths of 2_1^+ states, $B(E2;\ 2_1^+ \rightarrow 0_1^+)$ for even–even nuclei with $A = 60$–140 and 140–250 are shown in figures 2.14 and 2.15, respectively. These decay strengths are given in so-called Weisskopf units (W.u.), which are explained below. The correlation between $B(E2;\ 2_1^+ \rightarrow 0_1^+)$ and $E(2_1^+)$ is dramatized in figure 2.16. Note that this is a log–log plot that runs over two orders of magnitude on each axis.

If one adopts an axially symmetric rotor model then, cf. the expression equation (2.3) and chapter 3, the smaller the value of $E(2_1^+)$ the larger the moment of inertia and, by implication, the larger the deformation of the nucleus. Qualitatively, this provides a simple consistent picture.

Electromagnetic strengths in nuclei are quantified using Weisskopf units. These were introduced by Weisskopf [4]; he made the following key statements:

'We can estimate these matrix elements by the following exceedingly crude method. We assume that the radiation is caused by one single proton which moves independently within the nucleus, ... '

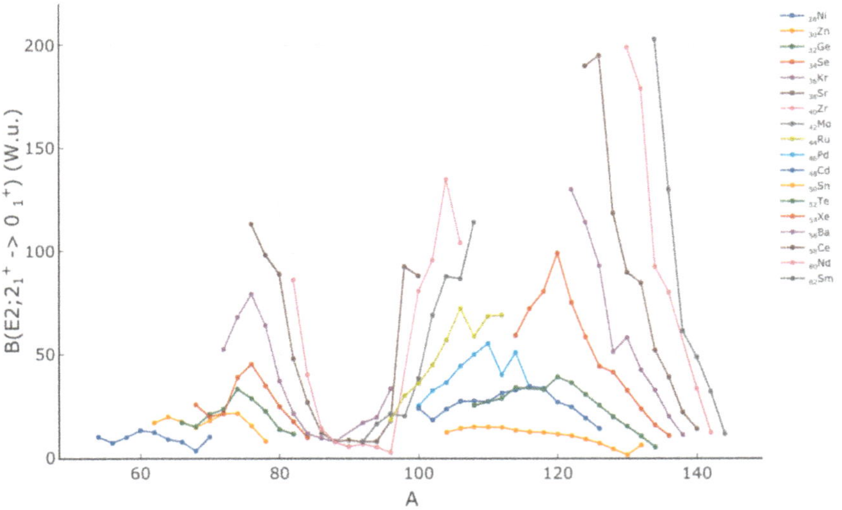

Figure 2.14. Map of $B(E2;\, 2_1^+ \to 0_1^+)$ for nuclei with $Z \geqslant 28$, $N \leqslant 82$. Isotopic sequences are connected by solid lines. Animation available at https://iopscience.iop.org/book/978-0-7503-2674-2.

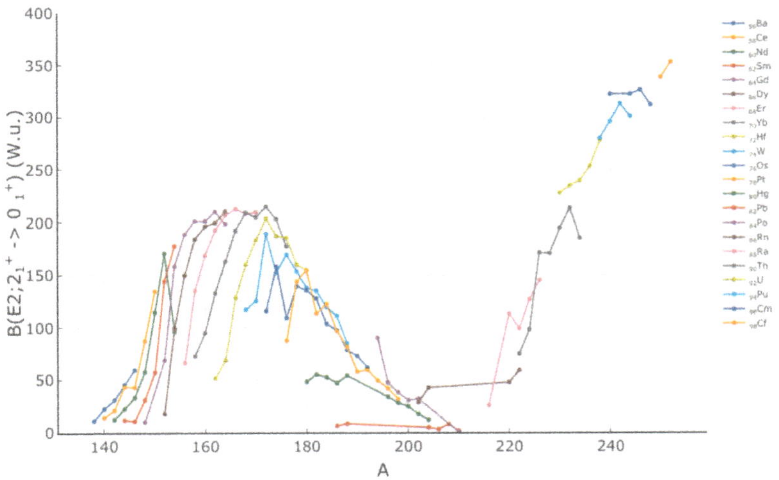

Figure 2.15. Map of $B(E2;\, 2_1^+ \to 0_1^+)$ for nuclei with $N \geqslant 82$. Animation available at https://iopscience.iop.org/book/978-0-7503-2674-2.

'The assumptions made in deriving these estimates are extremely crude and they should be applied to actual transitions with the greatest reservations. They are based upon an extreme application of the independent-particle model of the nucleus and it was assumed that a proton is responsible for the transition.'

For present purposes we give just the simplest practical relationships for employing Weisskopf estimates of electromagnetic strengths in nuclei. They relate the half-life and, by implication, the lifetime of the state to the electromagnetic decay

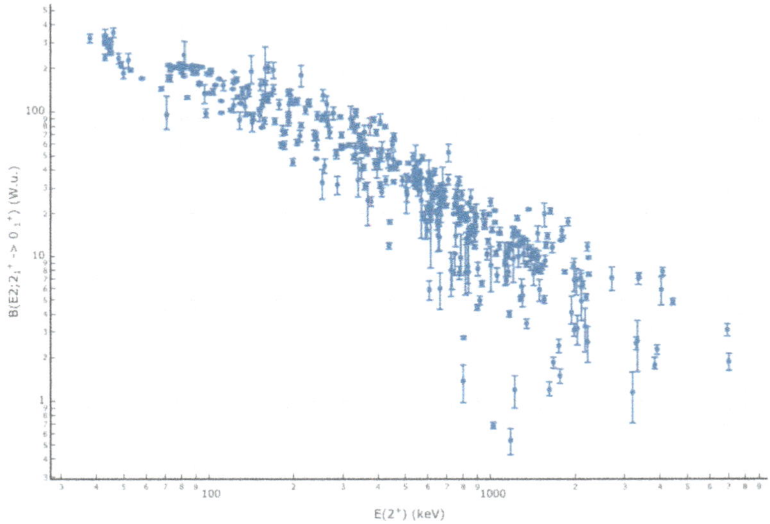

Figure 2.16. A log–log plot of $B(E2;\, 2_1^+ \rightarrow 0_1^+)$ versus $E(2_1^+)$ for all even–even nuclei. Animation available at https://iopscience.iop.org/book/978-0-7503-2674-2.

strength. They depend upon the branching fraction, b_γ for the transition of interest. Thus, the $E2$ decay strength in Weisskopf units is

$$B(E2) = \frac{9527}{E_\gamma^5 T_{1/2}(\gamma) A^{4/3}}, \tag{2.5}$$

where E_γ is the γ-ray transition energy in MeV and A is the mass number of the nucleus; $T_{1/2}(\gamma) = T_{1/2}/b_\gamma$, where $T_{1/2}$ is the half-life in picoseconds, and b_γ is the γ-decay branching fraction for the specific transition. Even if the state only de-excites by a single γ-ray transition, $b_\gamma < 1$ because of the internal conversion (and internal pair) decay processes. If the transition possesses mixed multipolarity, e.g. a $J = 2 \rightarrow J = 2$ transition always occurs by a mixture of $E2$ and $M1$ multipolarities, then the multipole mixing ratio, $\delta(E2/M1)$ must be measured (see chapter 6, figure 6.9). In such a circumstance,

$$b_\gamma(E2) = I_\gamma \frac{\delta^2}{(1 + \delta^2)} \frac{1}{1 + \alpha_{\mathrm{TOT}}}, \tag{2.6}$$

where α_{TOT} is the total internal conversion coefficient for the mixed $E2$ and $M1$ decay processes, summed over all atomic electron shells. (Details of internal pair decay are presented in chapter 6.) An alternative and commonly used unit for $E2$ decay strength is e^2b^2, where e is the unit of electric charge and $b := 1 \times 10^{-24}$ cm^2 and is called a 'barn'. For $E2$ decay, 1 W.u. $=5.940 \times 10^{-6} A^{4/3}$ e^2b^2. (Sometimes $E2$ decay strengths are expressed as e^2fm^4, where 1 b = 100 fm^2.)

All nuclear decays within a given nucleus involve the internal conversion process and all transitions with an energy above 1.022 MeV ($= 2m_ec^2$) involve the internal-pair formation process. Internal-conversion and internal-pair coefficients are discussed in chapter 6. A practical resource is the BrIcc program [5] maintained at Australian National University. Such decay modes are critical for the decay of excited 0^+ states where $J = 0 \rightarrow J = 0$ transitions[1] are involved. They are called electric monopole, $E0$ transitions.

At the heart of any Weisskopf estimate is an excited state half-life. These can be determined directly by observing intensity of radiations emitted from an ensemble of the decaying species as a function of time. Because half-lives are generally very short for states that undergo electromagnetic decay, timing is usually carried out electronically with a 'start' and 'stop' signal: start triggered by detection of a feeding radiation and stop by detection of a decaying radiation. However, for very short half-lives (<10 ps) even electronic means are too slow. Then the techniques of Doppler line broadening or Doppler energy shifts are used. Details are presented in chapter 6.

Electromagnetic transition strengths can also be measured using Coulomb excitation. The nuclear species of interest is either the target species or the beam species in a nuclear collision and is excited via the electromagnetic interaction. Depending on the charge and velocity of the beam, the excitation can be one-step or multi-step. Following excitation, the excited state 'relaxes' back to the ground state and this is observed via γ-ray emission. An example is presented in figure 2.2(a). The yield of intensities for the observed γ-rays, for a given beam (energy, Z) and target thickness, can be deconvoluted into a set of $E2$ matrix elements that describe the multiple paths of the excitation response of the nucleus. We discuss this and the role of other multipolarities in Coulomb excitation in chapter 6. The basic dependence of electromagnetic transition rates on energy and angular momentum, as observed in nuclei, is also explained in chapter 6.

2.2 A basic symmetry of nuclei: isospin

A basic and remarkable symmetry possessed by nuclei is revealed by excitation energies: excitations possess 'mirror' symmetry under exchange of protons and neutrons. This is illustrated in figure 2.17. The implication is that the nuclear force does not distinguish between protons and neutrons. The nuclear force dominates over Coulomb repulsion in nuclei, not only in the control of nuclear binding (total energy), but also in excitations (relative energy). The proton and neutron are said to be two manifestations of so-called isospin. This term has been coined to reflect the 'two-state' nature of nucleons, in analogy with 'spin-up' and 'spin-down' possessed by fermions. For the Reader with a more specialized interest we point to equation (2.89) in [6]: multiply by (i.e. take a direct product with) an isospin function, similar to the spin functions $\chi_\pm$. Thus, terms $\chi^{\mathrm{S}}_\pm \chi^{\mathrm{T}}_\pm$ would appear, where 'S' designates the spin degree of freedom and 'T' designates the isospin degree of freedom. The spin

[1] Note that electromagnetic radiation involves emission of light quanta, i.e. photons and they have a spin of 1; thus a $J = 0 \rightarrow J = 0$ transition cannot occur by single-photon emission. While two-photon emission can mediate this process in nuclei, by far the most probable process is internal conversion (and competing internal-pair formation at high energy).

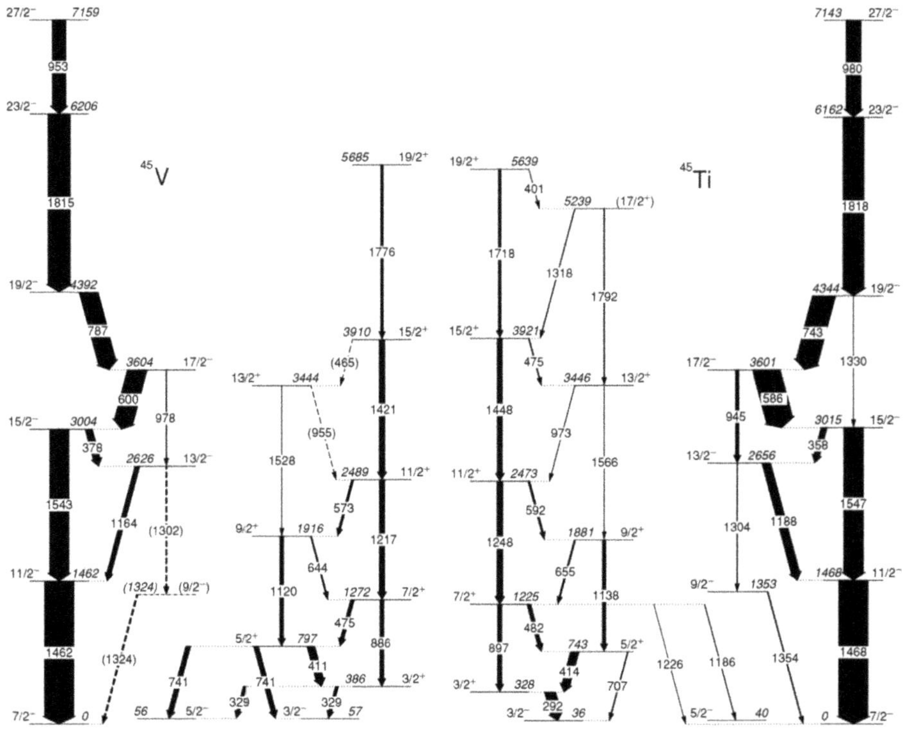

Figure 2.17. Decay patterns of excited states in the nuclei ${}^{45}_{23}V_{22}$ and ${}^{45}_{22}Ti_{23}$, a so-called 'mirror pair' of nuclei. The decay schemes are presented as reflected images to dramatize their independence with respect to whether the structures are $Z = 23$, $N = 22$ or $Z = 22$, $N = 23$. (The details of the structure involved are unimportant.) Reprinted with permission from [7]. Copyright (2006) by the American Physical Society.

degree of freedom couples to the spatial angular momentum degree of freedom (cf. equation (2.97) in [6]); but the isospin degree of freedom behaves as a 'spectator', hence the (near) independence of the spectra for these two nuclei with respect to whether isospin is up, ^{45}Ti or isospin is down, ^{45}V. (Note by convention, a neutron is assigned $T_z = +1/2$ and a proton is assigned $T_z = -1/2$.)

2.3 Nuclear data compilations and evaluations

Every practicing nuclear scientist needs to be able to 'navigate' the database of nuclear properties. This database is maintained and evaluated by a dedicated team of individuals. Evaluations are carried out at timely intervals to ensure that an up-to-date view of nuclear data is maintained. The National Nuclear Data Center maintains the Evaluated Nuclear Structure Data File, ENSDF and the Unevaluated Nuclear Data Lists, XUNDL. It also publishes evaluations, by A chain, at regular intervals per A chain, in the journal 'Nuclear Data Sheets'. Specialized data evaluations appear variously in journals of record, e.g. 'Atomic Data and Nuclear Data Tables' and 'Chinese Physics C'; and review articles that summarize special topics appear in 'Physics Reports', 'Progress in Particle and Nuclear Physics', 'Reports on Progress in Physics' and 'Reviews of Modern Physics'. These data can be accessed online at NNDC, and at AMDC.

2.4 Exercises

The relative challenge and time investment associated with the exercises is indicated from * to ***. Exercises marked with 'E' require interactive access to figures or watching included video exercises and will only have such functionality in the e-book version.

2.1. In figure 2.12 the line relating r_6 to r_4 corresponds to the empirical relationship $E_I = aI(I + 1) + bI^2(I + 1)^2$. Obtain an expression of the form $r_6 = cr_4 + d$. **

2-2. Figure 2.4 presents a conversion electron spectrum for the 126 keV transition in ^{160}Er populated via the beta decay of ^{160}Tm. Suppose that the assignment of the electron lines had not yet been made to the indicated transition in ^{160}Er. Assuming that the electron energies could be measured to a precision of 0.01 keV, how could the transition be definitely identified as occurring in a nucleus with $Z = 68$ (Er)? *

2-3. Use the BrIcc code to make α_K and K/L plots for:
   (a) $E_\gamma = 100$ keV as a function of Z for E1, M1, E2 [$Z = 20, 30, \ldots, 100$]. *
   (b) $Z = 68$ as a function of energy for E1, M1, E2 [$E_\gamma = 80, 90, 100, 120, 150, 200$ keV]. *

2-4. Using ENSDF, tabulate all of the unnatural-parity states for ^{152}Sm, i.e. states with spin-parity $0^-, 1^+, 2^-, 3^+, \cdots$ below 2 MeV. **

2-5. Figure 2.12 explores the relationship between r_6 and r_4 which are defined in equation (2.4) for nuclei which approximate closely to an ideal rotor.
   (a) Extend this study by defining r_8 by analogy with r_4 and r_6. *
   (b) Obtain the rotor limit value for r_8. *
   (c) Look up data using ENSDF for even–even rare earth and actinide nuclei around ^{172}Yb and ^{242}Pu, and make a plot of r_8 versus r_2. **

2-6. To obtain high precision on ratios of level energies as shown in figure 2.12, it may be necessary to perform a correction to the gamma-ray energies observed in experiment as gamma-ray emission is a two-body process where the emitting nucleus recoils so some small part of the transition energy is taken up by the recoil.
   (a) Given that the photon's energy and momentum are related by the standard equation, $E = pc$, make use of conservation of energy and momentum to derive a formula for the nuclear recoil energy as a function of the gamma-ray energy, E_γ and atomic mass of the recoiling nucleus, A. **
   (b) Estimate the magnitude of the recoil correction for $A = 240$ and $E_\gamma = 150$ keV. *
   (c) Estimate the magnitude of the recoil correction for $A = 12$ and $E_\gamma = 3213.79$, 4438.94 keV. Note, these are the energies of the cascade of γ-rays de-exciting the Hoyle state in ^{12}C. Thus, give your estimate of the excitation energy of the Hoyle state and compare with the value given in ENSDF. *

2-7. Figure 2.16 presents a log–log plot of B(E2;$2_1^+ \rightarrow 0_1^+$) as a function of the 2_1^+ excitation energy. Explore this interactive figure, which will indicate the associated isotope for each data point on the figure, and locate the outliers to the general trend. (Hint: there are several near the bottom of the figure.) Identify if there is any common feature to these outliers and interpret this based on the discussion of such nuclei elsewhere in the book. *** **E**

2-8. Derive the relation between half-life and mean lifetime. *

2-9. Using equation (2.5), make a table of $T_{1/2}(\gamma)$ values in seconds for a $B(E2)$ of 1 W.u. and $A = 10$, 100; $E_\gamma = 0.01$, 0.1, 1.0, 10 MeV. *

2-10. For the cases in exercise 2-9, use BrIcc to obtain $T_{1/2}$ values. [For $A = 10$, assume $Z = 5$; for $A = 100$, assume $Z = 40$.] *

2-11. For those reading the e-book, explore the set of video-based exercises presented in figure 2.18–2.26. These videos are short self-contained explanations of different aspects of the nuclear structure of ^{152}Sm and contain short exercises intended to encourage engagement with the nuclear structure data in the ENSDF database and to draw on many of

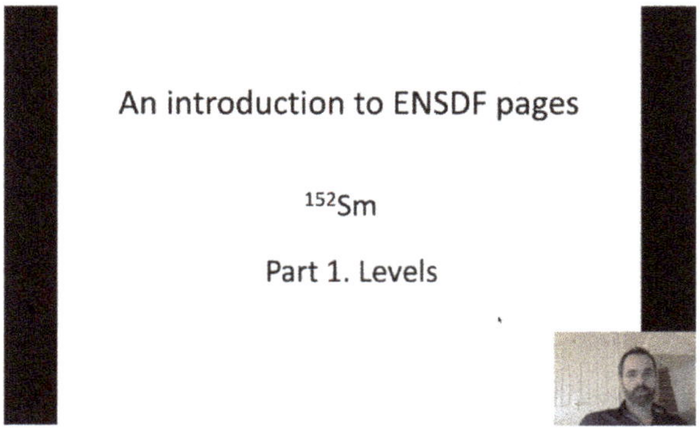

Figure 2.18. Video exercise on ^{152}Sm: Part 1. Levels. Video available at https://iopscience.iop.org/book/978-0-7503-2674-2.

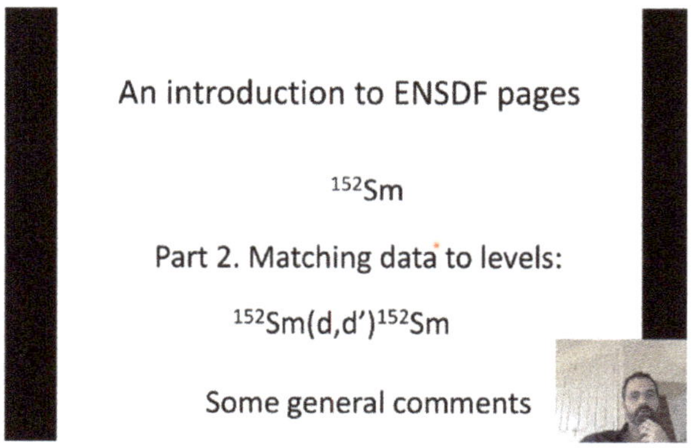

Figure 2.19. Video exercise on ^{152}Sm: Part 2. Matching data to levels. Video available at https://iopscience.iop.org/book/978-0-7503-2674-2.

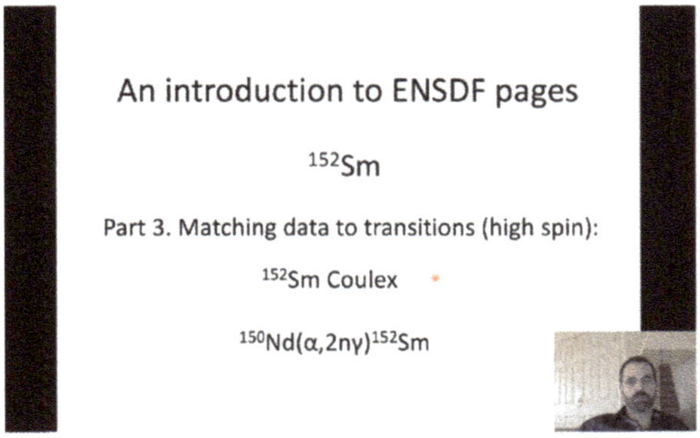

Figure 2.20. Video exercise on ^{152}Sm: Part 3. Matching data to transitions (high spin). Video available at https://iopscience.iop.org/book/978-0-7503-2674-2.

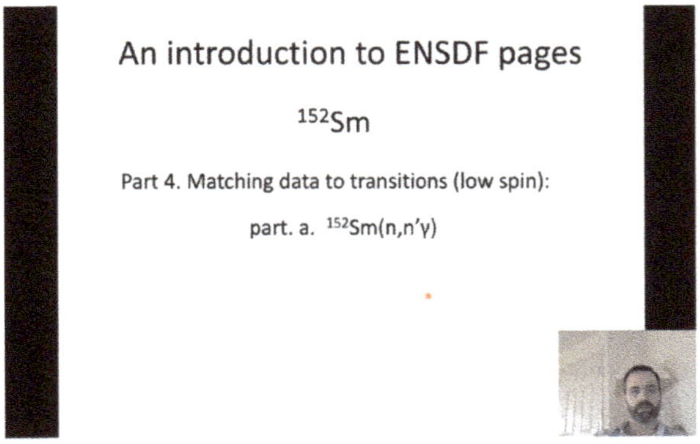

Figure 2.21. Video exercise on ^{152}Sm: Part 4. Matching data to transitions (low spin). Video available at https://iopscience.iop.org/book/978-0-7503-2674-2.

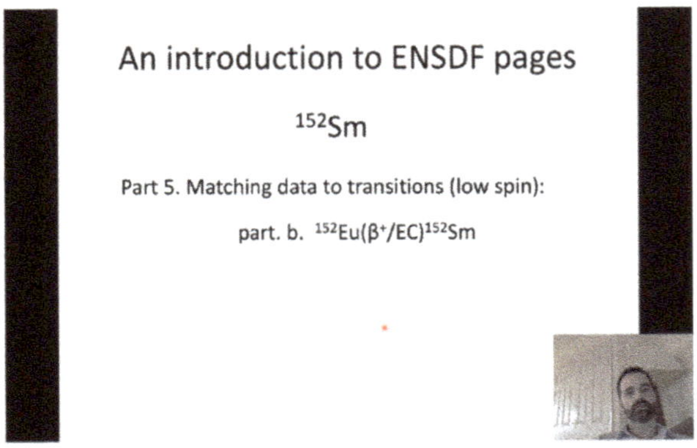

Figure 2.22. Video exercise on ^{152}Sm: Part 5. ^{152}Eu(β^{+}/EC)^{152}Sm. Video available at https://iopscience.iop.org/book/978-0-7503-2674-2.

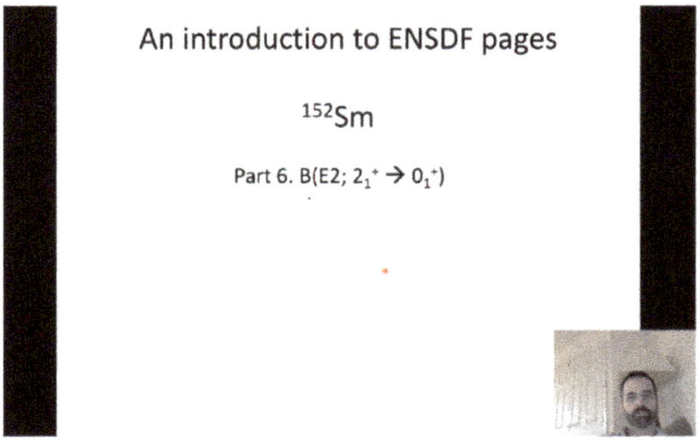

Figure 2.23. Video exercise on ^{152}Sm: Part 6. $B(E2;\ 2_1^+ \rightarrow 0_1^+)$. Video available at https://iopscience.iop.org/book/978-0-7503-2674-2.

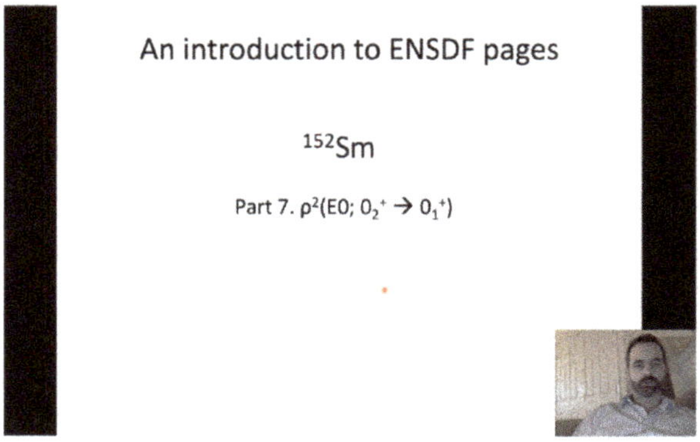

Figure 2.24. Video exercise on ^{152}Sm: Part 7. $\rho^2(E0;\ 0_2^+ \rightarrow 0_1^+)$. Video available at https://iopscience.iop.org/book/978-0-7503-2674-2.

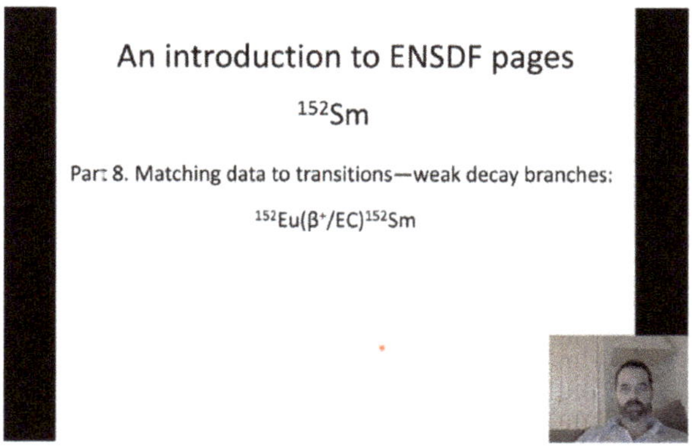

Figure 2.25. Video exercise on ^{152}Sm: Part 8. Matching data to transitions—weak decay branches. Video available at https://iopscience.iop.org/book/978-0-7503-2674-2.

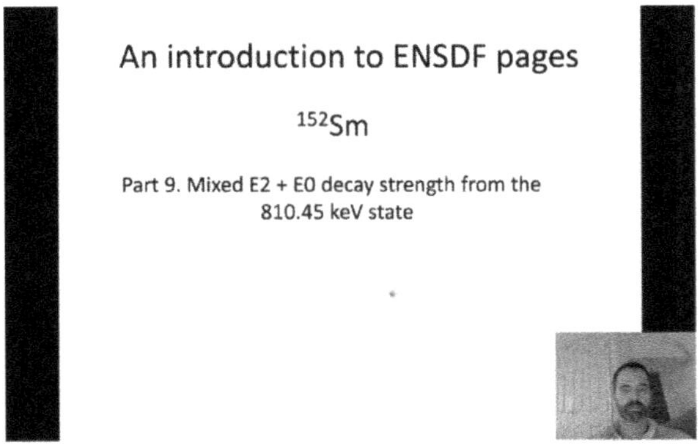

Figure 2.26. Video exercise on ^{152}Sm: Part 9. Mixed $E2 + E0$ decay strength from the 810.45 keV state. Video available at https://iopscience.iop.org/book/978-0-7503-2674-2.

the concepts and equations presented in this chapter. The powerpoints presented in the short videos can also be accessed here: [https://iopscience.iop.org/book/978-0-7503-2674-2]. *** **E**

References

[1] Heyde K and Wood J L 2019 *Quantum Mechanics for Nuclear Structure, Volume 1: A primer* (Bristol: IOP Publishing) https://iopscience.iop.org/book/978-0-7503-2179-2

[2] Smith J K, MacLean A D, Ashfield W, Chester A, Garnsworthy A B and Svensson C E 2019 Gamma–gamma angular correlation analysis techniques with the GRIFFIN spectrometer *Nucl. Instrum. Meth.* A **922** 47

[3] Taniuchi R *et al.* 2019 ^{78}Ni revealed as a doubly-magic stronghold against nuclear deformation *Nature* **569** 53

[4] Weisskopf V F 1951 Radiative transition probabilities in nuclei *Phys. Rev.* **83** 1073

[5] Kibédi T, Burrows T W, Trzhaskovskaya M B, Davidson P M and Nestor C W 2008 Evaluation of theoretical conversion coefficients using BrIcc *Nucl. Instrum. Meth.* A **589** 202

[6] Heyde K and Wood J L 2020 *Quantum Mechanics for Nuclear Structure, Volume 2: An intermediate level view* (Bristol: IOP Publishing) https://iopscience.iop.org/book/978-0-7503-2171-6

[7] Bentley M A *et al.* 2006 High-spin spectroscopy of natural and unnatural parity states in the mirror pair ^{45}V/^{45}Ti *Phys. Rev.* C **73** 024304

Chapter 3

Nuclear deformation and rotations

The dominance of rotational behavior patterns is emphasized. The peculiar nature of 'rotation' in quantum mechanical many-body systems is noted. The simple nature of models of rotation is introduced. The dominance of deformation in nuclei is stressed; and the occurrence of shape coexistence in nuclei is noted. Independent-particle motion in a deformed mean field, the Nilsson model, is described. A concise survey of the manifestation of rotational bands, their energies and E2 properties, is the chosen theme for the chapter. Some features of non-rotational states in deformed nuclei are provided. The K quantum number is used as the key quantum number for discussing rotational bands. K isomerism is sketched. Moments of inertia and a caution about the concept is addressed.

Concepts: quantum mechanical collective rotation, rotor model, axial and reflection symmetry, K quantum number, rotational eigenvectors, Clebsch–Gordan coefficients, Wigner–Eckart theorem, intrinsic quadrupole moment, intrinsic excitations, Nilsson model, particle-rotor model, decoupling, rotation alignment, gamma band, octupole band, broken-pair excitations, K isomerism, band crossing, band mixing, backbending, moments of inertia, superdeformation.

By far the majority of nuclei are deformed. A finite many-body quantum system that is deformed exhibits quantized rotations at a fundamental level. The deformation results in an infinite number of orientations, all with equal energies. Consequently, the quantum states of such systems are linear combinations of all of the orientations[1]. This associates collective rotations with the system and these rotations are quantized with

[1] A quantum system with a broken symmetry is said to possess a 'Goldstone' mode. The consequence of the broken symmetry is that the system possesses an infinite number of 'orientations of the axis' describing the symmetry breaking; these orientation configurations for the system are all degenerate in energy. When a quantum system possesses degenerate configurations, they generally mix and produce a lowest energy eigenstate. For the breaking of rotational symmetry in physical space, the lowest state is usually a state of lowest possible angular momentum. The literature on the Goldstone (or Anderson–Nambu–Goldstone) mode is not 'easy reading'. For the adventurous Reader, we suggest the invited comment in [1].

doi:10.1088/978-0-7503-2674-2ch3

states of well-defined energy and angular momentum. Such behavior is also associated with molecules. (We remind the Reader here that a mathematically expressed quantum state formalizes our lack of precise information for such systems. With respect to a deformed finite quantum system, a precise knowledge of orientation is not achievable.)

We choose to treat the nuclear data that are characteristic of deformed nuclei as a first step towards an organizational scheme because it is by far the dominant behavior in nuclei. Such behavior even occurs at low excitation energy in nuclei with spherical (non-deformed) ground states. This is termed 'shape coexistence' and is a further primary organizational scheme adopted in this work. Shape coexistence is introduced in chapter 4.

A highly attractive feature of quantized rotations is that they can be described using simple models. These models possess quantum mechanical properties that closely approximate phenomenology. Further, deviations from the simple model properties are straightforwardly accommodated by lowest-order corrections to the model, sometimes to a high level of precision. Details of model deviations from data and model corrections will be presented as the phenomenology dictates.

The handling of rotational data for deformed odd (and odd–odd) nuclei, as an extension of data for even–even nuclei, is straightforward and so all deformed nuclei are handled in this chapter. To make this step, the treatment of independent-particle motion in a deformed mean field must be made. This is called the Nilsson model [2]. The treatment subsumes the historical shell model, which at the present level is simply the spherical limit[2] of the Nilsson model. Deformed nuclei exhibit proton pairing correlations and neutron pairing correlations and these are briefly noted.

In the narrative on deformed nuclei we refer frequently to the nuclear 'shape'. However, the shape of a nucleus is not observable; only moments of the shape are observable. This is similar to the view of the nuclear density distribution, as depicted in figures 1.2 (inset) and 1.3, wherein the model expression, equation (1.1) is fitted to elastic scattering data such as shown in figure 1.2. The observables are the probabilities of detecting electrons at the various scattering angles with respect to the beam. The lowest-order moment of the density distribution is $\langle r^2 \rangle$, and this is an observable (the second-order moment would be $\langle r^4 \rangle$). These are called monopole moments. Extending this idea to nuclear shapes, the next-order shape moment after the monopole moment is the quadrupole moment. Nuclei do not possess dipole moments: this is a fundamental property of all finite many-body quantum systems, e.g. electrons, neutrons, atoms, molecules; in quantum mechanical terms, a moment is an expectation value, and the dipole moment operator is **r**—the average of **r** over physical space is from $-\infty$ to $+\infty$ and this is zero[3]. As the narrative evolves, particular attention is given to quadrupole moments (and higher shape moments).

[2] In the spherical limit, closed shells in nuclei usually are easy to discern; but the structures still involve correlations and these will need a framework of systematics in order to be understood.

[3] Beyond the domain of finite many-body quantum systems, at the level of the structure of space–time, searches are being made for an electric dipole moment (EDM) in these systems. If an EDM is observed, it would imply that space–time has a 'broken' inversion symmetry. Indeed, evidence from rare particle decays suggests this; but the topic is beyond the focus of the present narrative. The advanced Reader will find a useful introduction to EDM searches and their motivational basis in, e.g. [3].

But, one more caution is needed: these moments are defined in the body frame of reference or 'intrinsic' frame, and observations are made in the laboratory frame of reference. We address this in section 3.1.

3.1 The rotor model for nuclei: $K = 0$ bands

A deformed nucleus can be viewed, for the purpose of introducing the rotor model, as depicted in figure 3.1. The leading nuclear property that must be modeled is the intrinsic symmetry of the deformed system. We limit the treatment at this stage to the simplest possible deformation: a non-spherical shape with an axis of rotational symmetry, and a plane of reflection symmetry at right angles to this axis. There is an infinite variety of possible shapes, but we have in mind a quadrupole deformation. We will bias the treatment by always depicting prolate spheroidal deformation (positive quadrupole moments) because empirically this is realized in most nuclei (cf. figure 1.14). Oblate deformation results in a few differences beyond the sign of the quadrupole moment and these are handled where there is evidence for such deformation in nuclei. We emphasize that a quadrupole moment is only a single,

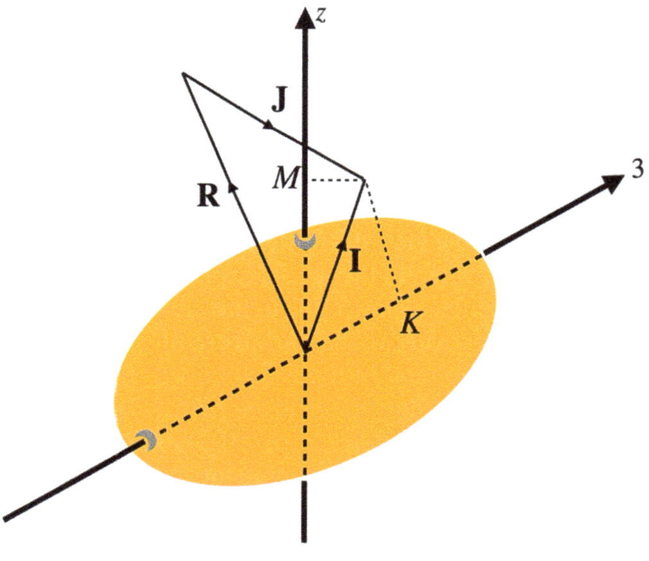

Figure 3.1. An axially symmetric rotor model viewed from a schematic perspective. The model possesses total spin **I**, intrinsic spin **J**, rotational angular momentum **R**, which are related by $\mathbf{I} = \mathbf{R} + \mathbf{J}$, and the Hamiltonian is given by equation (3.1) if $\mathbf{J} = 0$ and by equation (3.13) for non-zero **J**. When this model is quantized, the quantum number I, which labels the total spin of the system, together with the quantum numbers $I_3 := K$ and $I_z := M$ emerge (see text). The body-fixed symmetry axis is the 3-axis and the component of the total spin along this axis, labeled by K, is characteristic of the quantum mechanics of the system. The 3-axis and the z-axis (laboratory axis of directional quantization) are not in a fixed orientation with respect to each other (cf. figure 3.3). Quantum mechanically, rotations are restricted to directions perpendicular to the 3-axis and so K remains fixed for increasing values of **R**, i.e. K is a characteristic of a rotational band. Note that besides the axis of rotational symmetry, there is a plane of reflection symmetry at right angles to the 3-axis, passing through the centre of mass of the body. This plays a key role in the structure of the quantum mechanical state vectors, cf. equation (3.5).

linearly independent element in a basis, e.g. a spherical harmonic in an expansion of the mass distribution of a finite body[4].

For even–even nuclei, the adaptation of the rotor model is simplified because the ground state has $J = 0$. There are excited states in even–even nuclei that involve rotations built on intrinsic excitations with non-zero spin: these states will be handled along with odd-mass nuclei, following the introduction of the Nilsson model.

The Hamiltonian for the symmetric rotor is

$$H = \frac{I^2}{2\mathcal{J}}, \tag{3.1}$$

where I is the total angular momentum operator for the system and $\mathcal{J}$ is called the moment of inertia. The quantum mechanical basis states are the familiar states from the quantum theory of spin-angular momentum. However, there is an added 'dimension' to quantum rotors, namely they possess two independent axes of quantization: one fixed in the body frame and one in the laboratory frame. This leads to some subtle features of the quantum rotor model that lie beyond the quantization of angular momentum. This pertains to the so-called 'cone of indeterminacy', depicted in figure 3.2. The classical view of a symmetric rotor is that it can stably rotate about an axis at right angles to the symmetry axis. A quantum rotor possesses an uncertainty, which results in the classical view being 'too sharp'. Thus, the set of basis states needs two quantum numbers to describe directional quantization. These are shown in figure 3.1. This does not imply that

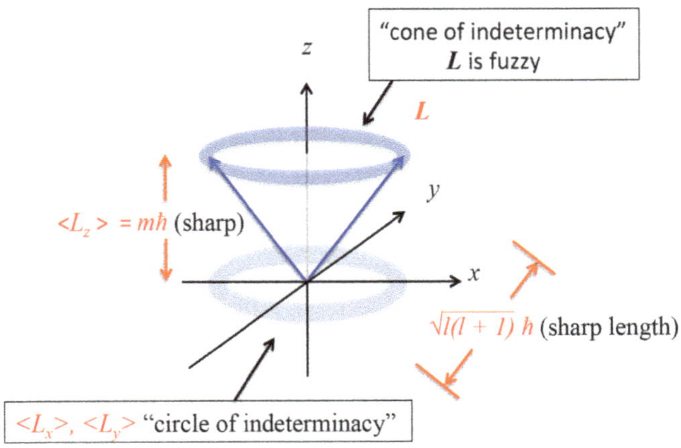

Figure 3.2. The 'cone of indeterminacy' for angular momentum in quantum mechanics. It is impossible to define more than one directional component of angular momentum, i.e. only one directional component is 'sharp'. Thus, in physical space, the other two (Cartesian) components can only be defined to within a circle. This results in the useful depiction of the quantum mechanical uncertainty using a cone of height m and side $\sqrt{l(l+1)}$, in units of $\hbar$.

[4] We draw the attention of the Reader to the GEOID, the gravitational equipotential of Earth.

there is simultaneous directional quantization along more than one reference frame axis (which, from the quantum theory of angular momentum, is impossible); it means that the symmetry axis of the nucleus possesses a 'hypercone' of indeterminacy with respect to the laboratory frame of reference as attempted in the depiction, figure 3.3.

The energy eigenvalues corresponding to equation (3.1) are (cf. equation 2.3)

$$E = \frac{\hbar^2 I(I+1)}{2\mathcal{J}}, \tag{3.2}$$

where $\hbar$ is the fundamental scale factor for quantum behavior. This relationship is often written as

$$E = AI(I+1), \tag{3.3}$$

where A is a model constant that must be fitted to data. This relationship can be directly viewed through a spectrum of γ-ray transition energies, as illustrated in figure 3.4. A γ-ray transition energy reflects an energy difference, i.e. the 'differential' of equation (3.3) with respect to spin, viz.

$$E_\gamma = \Delta E_{I,I-2} = A(4I-2). \tag{3.4}$$

Thus, the γ-ray energies in figure 3.4 would be equidistantly spaced (a linear function of I) if the nuclei, ^{242,244}Pu were simple rotors. Evidently, from the decrease in

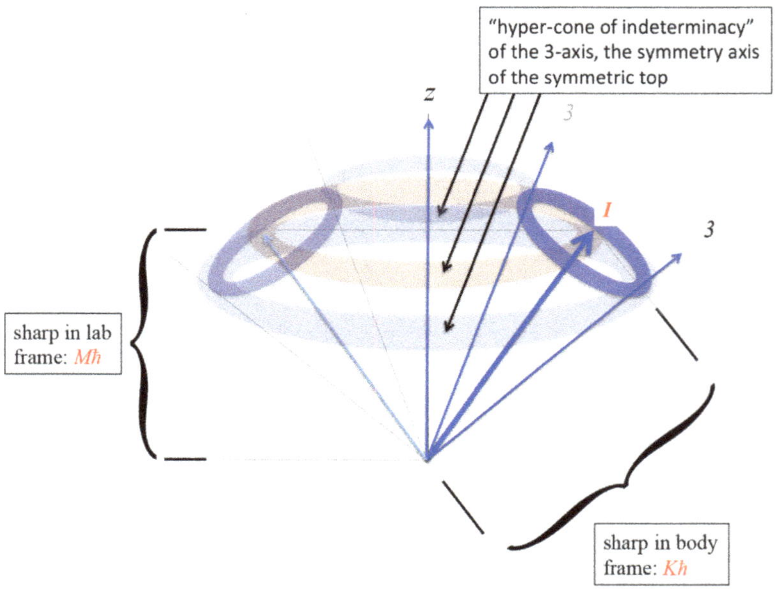

Figure 3.3. The 'hypercone of indeterminacy' for the quantum mechanical axially symmetric rotor. The body-fixed frame and the laboratory frame are not in a fixed relationship to each other: they are connected by the total spin I possessing a fixed projection $M\hbar$ on the laboratory frame z-axis and a fixed projection $K\hbar$ on the body frame 3-axis. In consequence, the 3-axis has an uncertainty that involves two cones of uncertainty from which the figure endeavors to depict the full degree of uncertainty possessed by the quantum axially symmetric rotor.

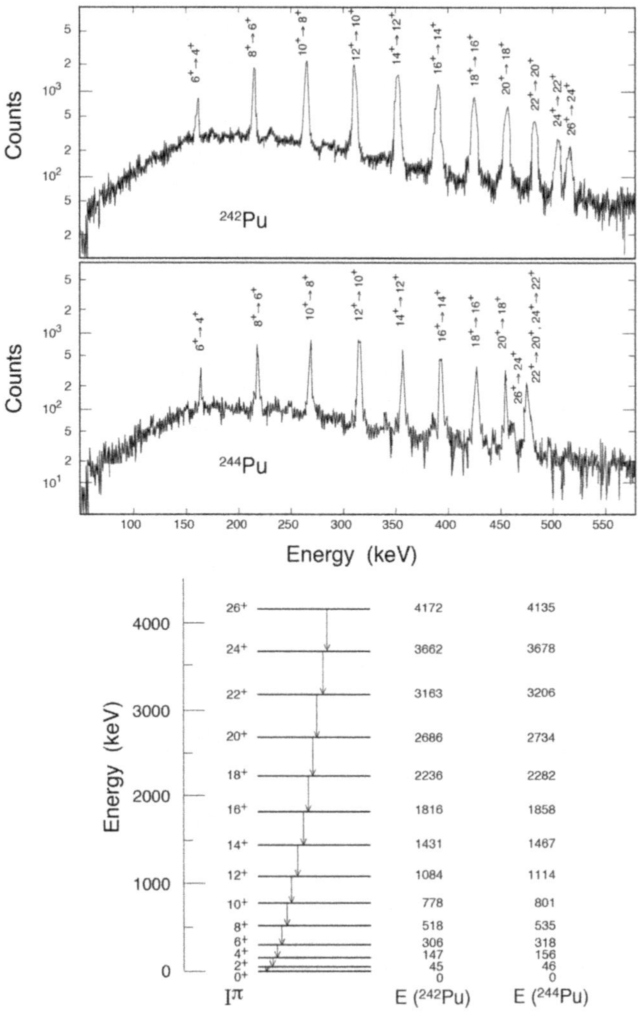

Figure 3.4. Spectrum of an axially symmetric rotor, as manifested in $^{242,244}Pu$ following Coulomb excitation. Details are discussed in the text. The figure is reproduced from [4], copyright (2010) World Scientific Publishing Company. The figure is a representation of original data appearing in [5].

spacing of the γ-ray energies with increasing spin, it can be inferred that the parameter A must be decreasing with increasing spin. Further, in figure 3.4 it is evident that above spin 22 in ^{244}Pu the spacing becomes irregular. These features of nuclear rotation are discussed in section 3.4.3.

There are further subtleties to quantum rotors: if $\mathcal{J}$ is small, the energy scale of quantization may be very big, i.e. $\hbar^2/2\mathcal{J}$ may be large. This occurs in atoms and is the reason that one never discusses collective rotations in atoms. There are other reasons such as atoms possessing rotational symmetry and so, from their quantum mechanical nature, 'rotation' has no meaning (at the most fundamental level, atoms do not possess a Goldstone mode). This is the reason that, for the quantum symmetric rotor, rotations about the symmetry axis are not discussed. Later, rotations about such an axis are introduced in nuclei when this axis is not rotationally symmetric.

The allowed values of I for ground-state rotational bands of even–even nuclei, which have $J = 0$ ground states, are $I = 0, 2, 4, 6, 8, \ldots$. The reason for only even-spin values follows from the symmetry of the eigenstates: one sees from figure 3.1 that the two states $|IMK\rangle$ and $|IM, -K\rangle$ are degenerate in energy (there is reflection symmetry). Thus, the correct ground state is

$$|IM\rangle = \frac{\{|IMK\rangle + (-1)^{I+K}|IM, -K\rangle\}}{\sqrt{2}}, \tag{3.5}$$

where the relative phase factor expresses the reflection symmetry of the angular momentum states; but for $K = 0$ this only allows even values of I because $|IMK\rangle$ and $|IM, -K\rangle$ become indistinguishable. The energies manifested in ground-state bands of even–even nuclei accurately reflect an even-spin only sequence, albeit the spins are established from experiment, and they match this. (We remind the Reader that, in the mathematical formulation of quantum mechanical systems, we cannot write equations that imply knowledge that is unattainable. Often, differences between quantum mechanics and classical mechanics can be essentially incomprehensible.)

Electromagnetic properties for ground-state bands of even–even nuclei follow from the axially symmetric rotor model. The 'even-spin only' states confine electromagnetic transitions to $E2$ multipolarity. Matrix elements follow from the Wigner–Eckart theorem[5], viz.

$$\langle IMK = 0|T(E2)|I - 2, MK = 0\rangle = (2I + 1)^{1/2}\left(\frac{5}{16\pi}\right)^{1/2}\langle I020|I - 2, 0\rangle eQ_0, \tag{3.6}$$

where $\langle I020|I - 2, 0\rangle$ is a Clebsch–Gordan coefficient and Q_0 embodies the so-called reduced matrix element for the $E2$ operator and is a parameter for the model, e is the fundamental unit of electrical charge.

The resulting $B(E2)$ values for the ground-state band of an even–even nucleus that characterize the quantum symmetric rotor are given by

$$\frac{B_{I,I-2}}{B_{20}} = \frac{15I(I - 1)}{2(2I - 1)(2I + 1)}, \tag{3.7}$$

where

$$B_{I,I-2} := B(E2; I \to I - 2) = \frac{\langle IMK = 0|T(E2)|I - 2, MK = 0\rangle^2}{2I + 1}. \tag{3.8}$$

[5] It is important to note the role of the K quantum number here. The K quantum number is peculiar to the model (it does not have a 'microscopic' justification). It plays the key role in the 'reduction' of the matrix element in equation (3.6) to yield a reduced matrix element, which becomes the constant Q_0, the intrinsic quadrupole moment (a model-defined quantity): this quantity is a characteristic of the rotational band. Many ratios of measured electromagnetic properties then emerge, which are expressed in terms of just the total spin quantum number in simple algebraic relations.

This leads to the values: $I(B_{I,I-2}/B_{20}) = 4(1.43), 6(1.57), \ldots$. Figure 3.5 shows that, within experimental uncertainty, ground-state bands of deformed even–even nuclei are well described by the axially symmetric rotor model.

Electric quadrupole moments of states in the ground-state band of even–even nuclei likewise can be expressed in terms of the parameter Q_0. Again, matrix elements follow from the Wigner–Eckart theorem, viz.

$$\langle IMK = 0|T(E2)|IMK = 0\rangle = (2I + 1)^{1/2}\left(\frac{5}{16\pi}\right)^{1/2}\langle I020|I0\rangle eQ_0. \tag{3.9}$$

The resulting $Q(I)$ values for the ground-state band of an even–even nucleus that characterize the quantum symmetric rotor are given by

$$Q(I) = \left[-\frac{I}{2I + 3}\right]eQ_0. \tag{3.10}$$

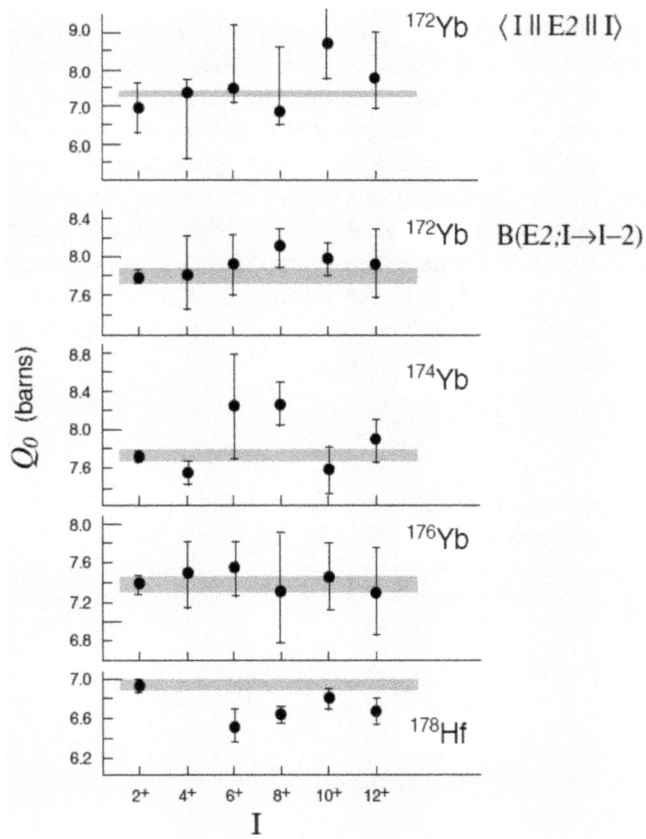

Figure 3.5. Electromagnetic transition and diagonal $E2$ matrix elements reduced to intrinsic quadrupole moment, Q_0 values, cf. equations (3.6) and (3.9). The shaded bands are the values of Q_0 deduced from the matrix element connecting the 2_1^+ and 0_1^+ states. The figure is reproduced from [4], copyright (2010) World Scientific Publishing Company.

Note that the quadrupole moment of the ground state ($I = 0$) is zero: this reflects the indeterminacy of a quantum rotor, i.e. it appears as a spherically symmetric body when the spin is zero. As the spin increases, $Q(I)/Q_0$ takes the values $-\frac{2}{7} = -0.29$, $-\frac{4}{12} = -0.33$, $-\frac{6}{15} = -0.40$, ..., and approaches a maximum value of −0.5. Thus, one never observes the full magnitude of the intrinsic quadrupole deformation. Note further the negative sign, i.e. a positive Q_0 (prolate spheroid in the body frame) appears as an oblate spheroid in the laboratory frame: for a naïve view of this, consider a rugby football spinning around a fixed axis. Data for diagonal $E2$ matrix elements (equation (3.9)) and quadrupole moments (equation (3.10)) are sparse and are rarely of high precision[6]. Some values are shown in figure 3.5.

3.2 Odd-mass nuclei, intrinsic excitations, and rotations

Odd-mass nuclei possess an unpaired nucleon. To a first approximation, unpaired nucleons are described by independent-particle degrees of freedom: these degrees of freedom are described by mean-field models, where the nucleon is confined in a potential that closely approximates the density distribution of the nucleons in the nucleus. The essential contributions to independent-particle energies in spherical and deformed nuclei are depicted using diagrams shown in figures 3.6 and 3.7(a)–(c).

In considering the coupling of an unpaired particle to a rotor, it is useful to focus on the coupling of a single-j particle to an axially symmetric rotor. Although a deformed potential mixes configurations with different j values, high-j configurations are isolated by spin–orbit coupling (cf. figure 3.6) which produces a lowering of such a configuration so that it is a unique-parity configuration in a given shell. Note that parity for a nucleus is a good quantum number and so such high-j configurations have approximately pure j values. The model is called the axially symmetric particle-rotor model and the model Hamiltonian is

$$H = \frac{R^2}{2\mathcal{J}}, \tag{3.11}$$

where the rotational angular momentum, $\mathbf{R}$ is related to the total spin of the nucleus, $\mathbf{I}$ and the spin of the unpaired particle, $\mathbf{j}$ by (cf. figure 3.1)

$$\mathbf{R} = \mathbf{I} - \mathbf{j}. \tag{3.12}$$

Thus,

$$H = \frac{(\mathbf{I} - \mathbf{j}) \cdot (\mathbf{I} - \mathbf{j})}{2\mathcal{J}} \tag{3.13}$$

and expansion of the scalar product yields

$$H = \frac{I^2}{2\mathcal{J}} - \frac{2\mathbf{I} \cdot \mathbf{j}}{2\mathcal{J}} + \frac{j^2}{2\mathcal{J}}. \tag{3.14}$$

[6] The exception is quadrupole moments deduced from muonic x-ray data.

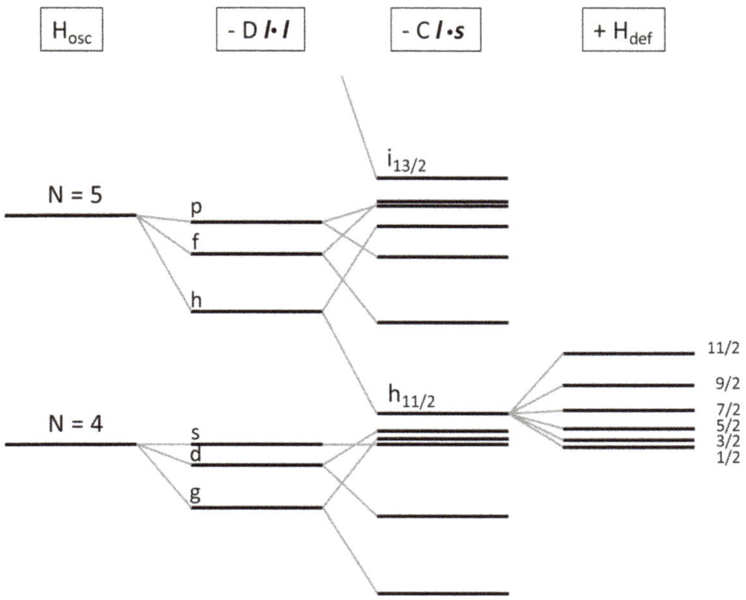

Figure 3.6. The hierarchy of quantized energy levels for independent-particle motion in nuclei, extending from spherical to spheroidal deformed mean fields. On the left-hand side is the prototype model harmonic oscillator potential: this is 'flattened' by the model term, $-Dl^2$ that effectively lowers high angular momentum orbitals. The spin–orbit splitting is described by the model term, $-C\mathbf{l} \cdot \mathbf{s}$. Finally, on the right-hand side is the spectrum (shown just for one of the j subshells) that one observes in a prolate spheroidally deformed mean field. In general, there is j-subshell mixing—see figure 3.7. The relative magnitudes of the energy scales are not in proportion to the physical scales and are only chosen to make the figure clearer. The angular momentum values that occur for the $N = 4$ and 5 oscillator shells, $l = 0, 2, 4$ (s, d, g) and $l = 1, 3, 5$ (p, f, h), respectively, are indicated. The spin–orbit splitting lowers the $j = l + 1/2$ configuration and raises the $j = l - 1/2$ configuration; thus, the $j = 5 + 1/2 = 11/2$ independent-particle state is lowered into the 'N = 4' shell. Parities of the orbitals are given by $(-1)^l$; thus, the $h_{11/2}$ orbital is a unique-parity orbital in the energy shell and does not mix with the $N = 4$ orbitals. More details on the Nilsson model approach to the description of unpaired nucleons are given in chapter 5.

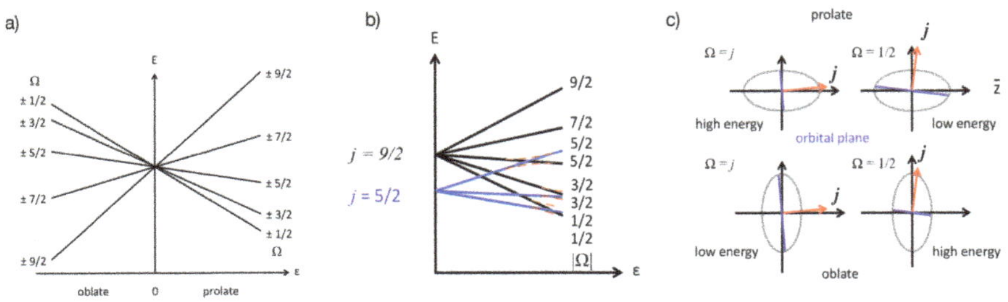

Figure 3.7. (a) States in a spheroidal potential ranging from oblate to prolate deformation. They are labeled with the quantum number $\Omega = j_3 = K$. (b) Configurations with identical $|\Omega|$ values and parities mix. Note that the energy eigenstates are linear combinations of $|+\Omega\rangle$ and $|-\Omega\rangle$, cf. equation (3.5), due to the plane of reflection symmetry of the model. This is the Nilsson model, further details of which are given in chapter 5. (c) A schematic view of the energy ordering of single-particle states in a spheroidal potential. The anisotropy of the harmonic oscillator potential results in 'narrow' confinement (high energy) or 'wide' confinement (low energy): this is termed 'deformation alignment' and the energy ordering is dictated by the Pauli principle. Each Nilsson state is twofold degenerate as expressed by $|\pm\Omega\rangle$.

The energy eigenvalues are

$$E = \langle IMKj|H|IMKJ\rangle = \frac{\hbar^2 I(I+1)}{2\mathcal{J}} - \frac{2\langle \mathbf{I}\cdot\mathbf{j}\rangle}{2\mathcal{J}} + \frac{\langle j^2\rangle}{2\mathcal{J}}, \tag{3.15}$$

where $\langle \mathbf{I}\cdot\mathbf{j}\rangle$ and $\langle j^2\rangle$ are quantum mechanical expectation values. This allows for the fact that j is not a good quantum number in a deformed mean-field. Rotational bands in an odd-mass nucleus are shown in figure 3.8(a); and identification of the intrinsic structure associated with the unpaired particle, using a so-called Nilsson diagram, is shown in figure 3.8(b) (details of Nilsson diagrams and the Nilsson

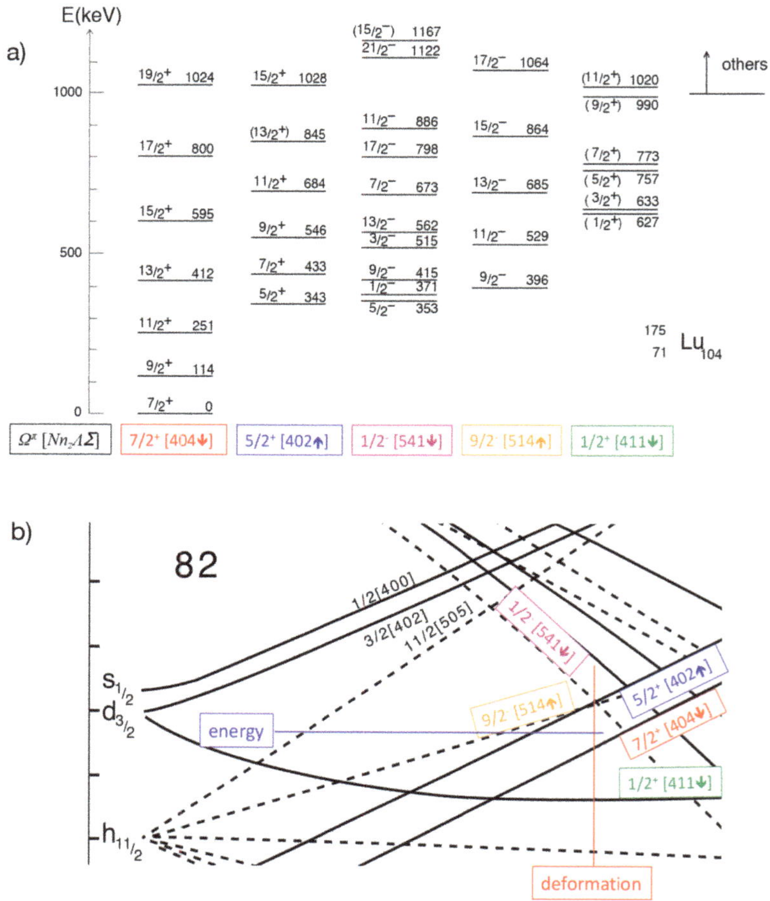

Figure 3.8. (a) and (b) An example of the Nilsson-plus-rotor model applied to the lowest energy rotational bands in ^{175}Lu. The location in the Nilsson diagram is fixed by the proton number ($Z = 71$) and by an estimate of the deformation characterized by the quadrupole moment of this nucleus: details are given in chapter 5; but it is evident that the deformation is narrowly defined by the observed band-head spins. Usually, band-head spin = Ω. The figure is adapted from [4], copyright (2010) World Scientific Publishing Company. Extended video-based exercises related to ^{175}Lu and the Nilsson model are found in exercise 3-6.

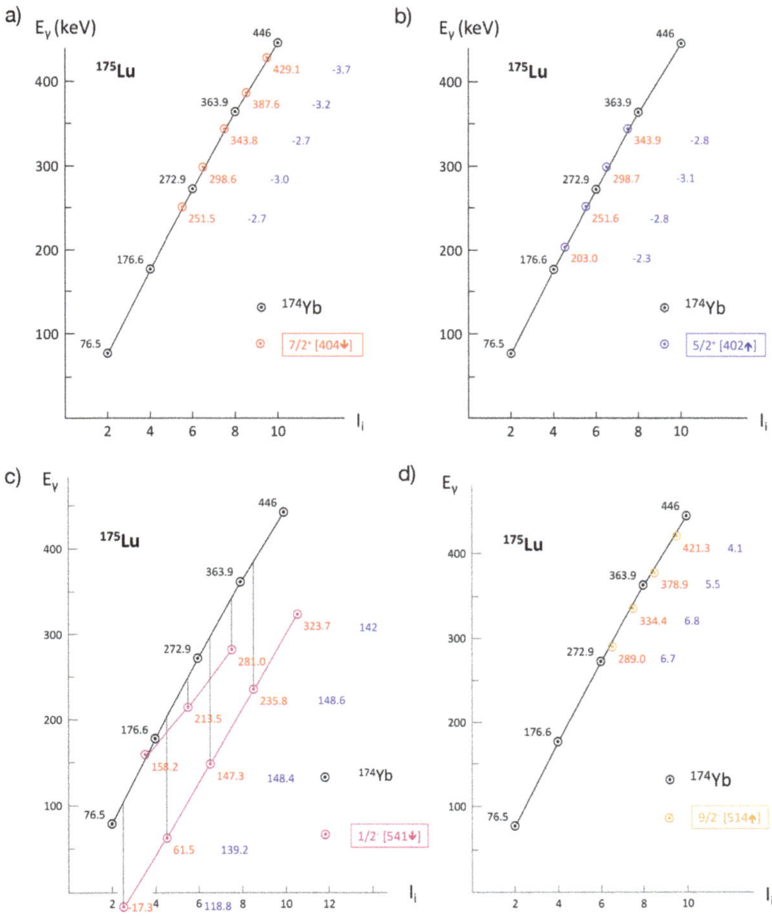

Figure 3.9. (a)–(d) Plots of γ-ray energy versus spin of the initial state, I_i, for transitions in the ^{175}Lu bands, compared to the neighboring even–even nucleus, ^{174}Yb. The γ-ray energies in keV are given in red and the interpolated energy differences are given in blue. Recognizing that transition energy is an energy difference for the rotor energy relationship, equation (3.15), it is evident that if the odd-mass profiles for such plots are parallel to the profile for the even–even 'core' nucleus, then the term $\langle \mathbf{I} \cdot \mathbf{j} \rangle$ has a simple constant contribution to the transition energy and the term $\langle j^2 \rangle$ does not change with I. Extended video-based exercises related to ^{175}Lu and the Nilsson model are found in exercise 3-6.

model are presented in chapter 5). A direct comparison between these bands and the ground-state rotational bands of neighboring even–even nuclei is shown in figure 3.9(a)–(d). Evidently, the terms $2\langle \mathbf{I} \cdot \mathbf{j} \rangle / 2\mathcal{J}$ and $\langle j^2 \rangle / 2\mathcal{J}$ do not play a major role for the odd-mass bands shown. This is discussed in some depth in the following. Noting that the figures plot γ-ray energies, which are energy *differences* with respect to equation (3.15), evidently these j-dependent terms are constant contributions for the bands shown.

Figure 3.10 shows the explanation of the $\mathbf{I} \cdot \mathbf{j}$ coupling in semi-classical terms, i.e. it is a simple issue of the degree of alignment of $\mathbf{j}$ with respect to $\mathbf{I}$. Figure 3.10 depicts two extreme coupling alignments:

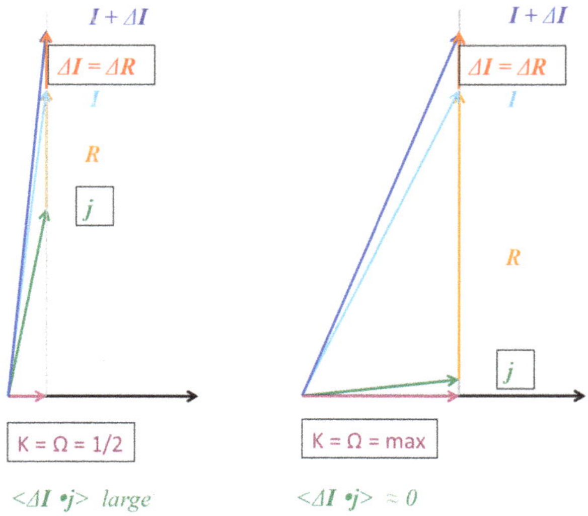

Figure 3.10. A semi-classical explanation of the effect of the $\langle \mathbf{I} \cdot \mathbf{j} \rangle$ term on rotational energies in nuclei. The details of the explanation are given in the text. Here, we note that the plots presented in figures 3.9(a)–(c) involve energy differences and they must be matched to vectorial differences of the $\langle \mathbf{I} \cdot \mathbf{j} \rangle$ term: recall, a small increment in the direction of a vector is at right angles to the vector, hence the large difference between the alignment shown on the left and that shown on the right.

1. On the right-hand side the single-particle spin, $\mathbf{j}$ is in maximum alignment along the symmetry axis, $\Omega = j$ and the rotational angular momentum, $\mathbf{R}$ is nearly at right angles to $\mathbf{j}$; hence, increments in the rotational angular momentum, $\Delta\mathbf{R}$ are nearly at right angles to $\mathbf{j}$ and thus, $\Delta\mathbf{R} \cdot \mathbf{j} = \Delta\mathbf{I} \cdot \mathbf{j} \approx 0$.
2. On the left-hand side the single-particle spin, $\mathbf{j}$ is in minimum alignment along the symmetry axis, $\Omega = 1/2$ and the rotational angular momentum, $\mathbf{R}$ is nearly parallel to $\mathbf{j}$; hence, increments in the rotational angular momentum, $\Delta\mathbf{R} = \Delta\mathbf{I}$ are nearly parallel to $\mathbf{j}$ and thus, $\Delta\mathbf{I} \cdot \mathbf{j} \approx$ maximum. In consequence, plots of $E_\gamma = \Delta E$ versus I_i reveal the effect of $\Delta\mathbf{I} \cdot \mathbf{j}$ and this is often either vanishingly small ($\Omega \approx j$) or, for $\Omega = 1/2$ it can be large.

A number of extensions to the basic rotor model, as applied to odd-mass nuclei can be accommodated within the above formalism. Figure 3.11 shows an E_γ versus I_{initial} plot for a $K = 1/2$ band. Such bands exhibit a 'signature' splitting which is evident to various degrees in odd-mass rotational bands, and is especially pronounced for $K = 1/2$ bands. There is a general formula for the energies of a $K = 1/2$ band, viz.

$$E = A[I(I+1) + a(-1)^{I+1/2}(I + 1/2)], \tag{3.16}$$

where a is a new model parameter called a 'decoupling' parameter. This formula is a consequence of the term $2\langle \mathbf{I} \cdot \mathbf{j} \rangle / \mathcal{J}$ when treated quantum mechanically. This is derived in chapter 5. Figure 3.12 illustrates the spin ordering, as a function of excitation energy, for such bands. Two examples of such bands are evident in figure 3.8(a).

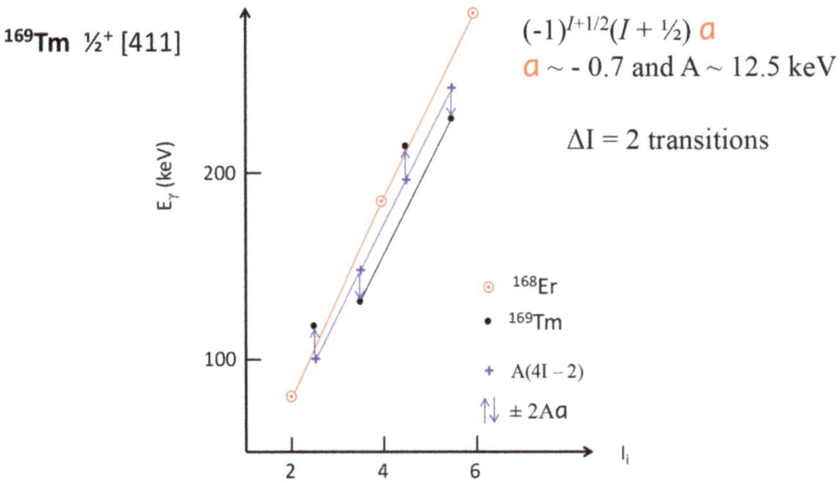

Figure 3.11. Effect of the $\langle \mathbf{I} \cdot \mathbf{j} \rangle$ term on a $K = 1/2$ band, illustrated for the $1/2^{+}[411]$ band in ^{169}Tm. The result is a so-called 'signature splitting'. The quantum mechanical treatment of the $\langle \mathbf{I} \cdot \mathbf{j} \rangle$ term leads to equation (3.16) (see text). For $\Delta I = 2$ transitions, the first term in equation (3.16) gives $A(4I - 2)$, shown for A ~12.5 keV; the second term in equation (3.16) gives $\pm 2Aa$, shown for $a \sim -0.7$.

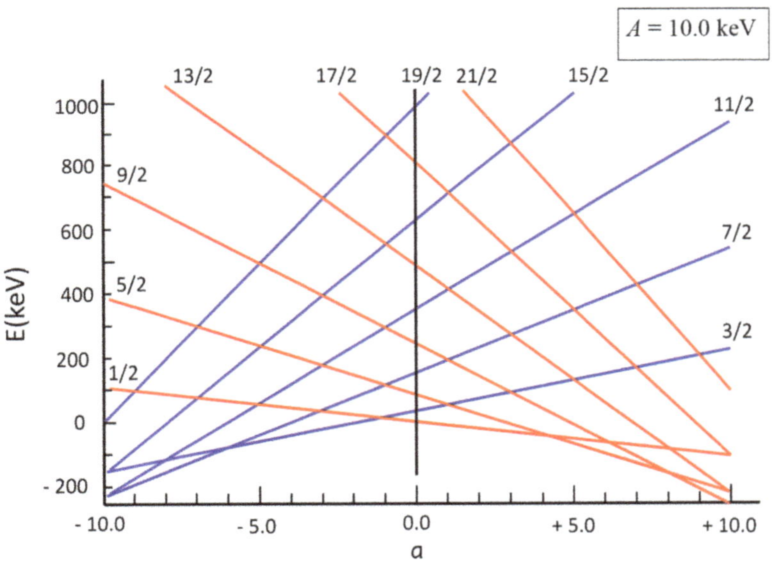

Figure 3.12. Depiction of spin sequences that are associated with equation (3.16) for different values of the model parameter a. The parameter A is set equal to 10.0 keV.

Electromagnetic properties for rotational bands in odd-mass nuclei follow from the axially symmetric rotor model in a manner identical to even–even nuclei. For the present, we confine the discussion to electromagnetic transitions with $E2$ multipolarity. Electromagnetic properties of ground-state bands in odd-mass nuclei can be directly studied by Coulomb excitation as shown in figure 3.13. Matrix elements follow from the Wigner–Eckart theorem, viz.

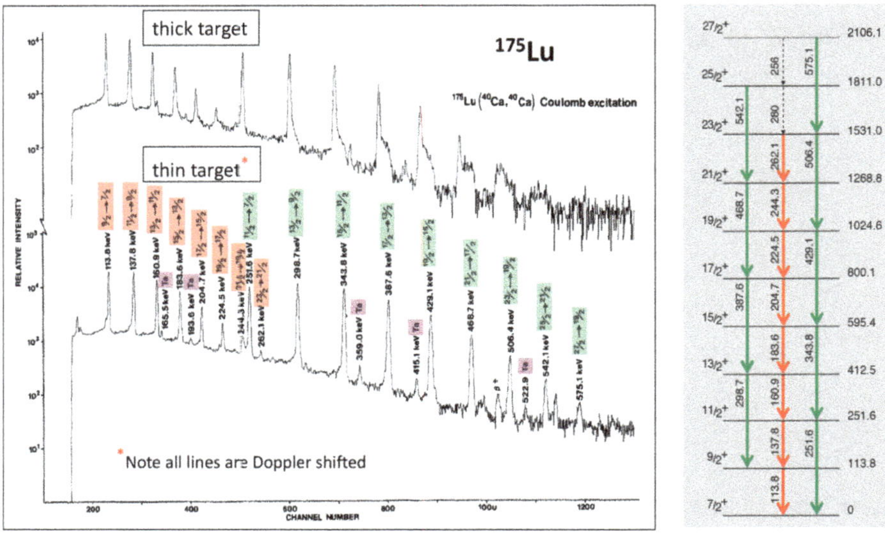

Figure 3.13. The γ-ray spectrum resulting from Coulomb excitation of a deformed odd-mass nucleus, ^{175}Lu (cf. figure 3.8). Spectra are shown for a 'thin target' and a 'thick target': the distinction is that in the spectrum from the former target, all of the γ-ray lines are Doppler shifted because the excited nuclei all recoil out of the target with the same recoil velocity; in the latter target, there is a slowing down of recoiling nuclei depending on where in the target they are excited, thus one observes a 'Doppler-broadened' γ-ray line shape. (The target material contained a small amount of ^{181}Ta as an impurity.) The inset shows the ground-state band of ^{175}Lu for convenient reference, cf. figure 3.8. (Reprinted from [6] with permission from Elsevier.)

$$\langle I_i MK|T(E2)|I_f MK\rangle = (2I_i + 1)^{1/2}\left(\frac{5}{16\pi}\right)^{1/2}\langle I_i K20|I_f K\rangle eQ_0, \tag{3.17}$$

where $\langle I_i K20|I_f K\rangle$ is a Clebsch–Gordan coefficient, Q_0 embodies the so-called reduced matrix element for the $E2$ operator and is a parameter for the model. Figures 3.14(a)–(c) show that, within experimental uncertainty, rotational bands of deformed odd-mass nuclei are well described by the axially symmetric rotor model. Figure 3.15 shows that variation of the intrinsic quadrupole moment model parameter, Q_0 exhibits an apparent monotonic trend as a function of mass.

Electric quadrupole moments of states in odd-mass nuclei likewise can be expressed in terms of the parameter Q_0. Again, matrix elements follow from the Wigner–Eckart theorem, viz.

$$\langle IMK|T(E2)|IMK\rangle = (2I + 1)^{1/2}\left(\frac{5}{16\pi}\right)^{1/2}\langle IK20|IK\rangle eQ_0. \tag{3.18}$$

The resulting $Q(I,K)$ values for an odd-mass nucleus that characterize the quantum symmetric rotor are given by

$$Q(I, K) = \frac{[3K^2 - I(I + 1)]eQ_0}{[(I + 1)(2I + 3)]}. \tag{3.19}$$

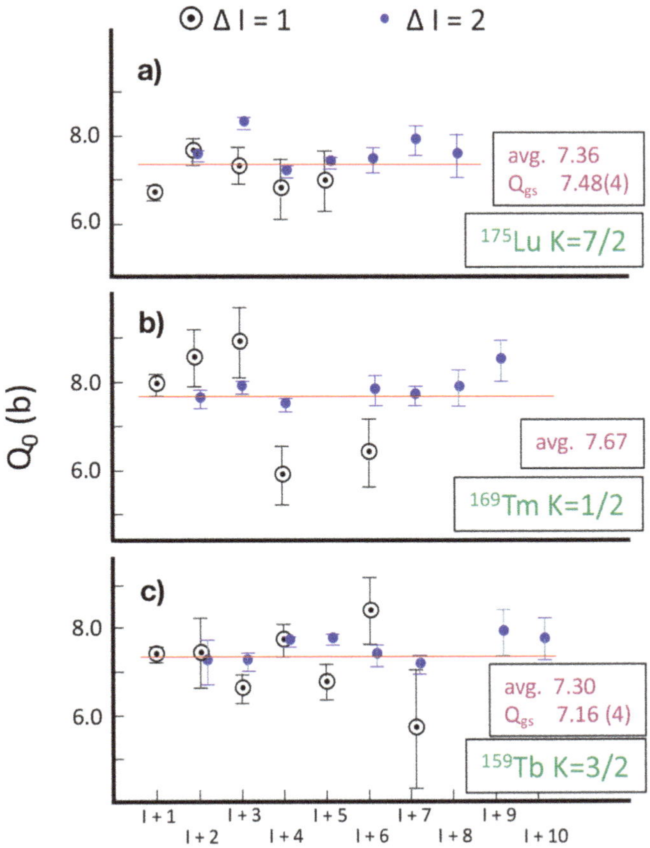

Figure 3.14. (a)–(c) Intrinsic quadrupole moments, Q_0 for selected odd-mass deformed nuclei as deduced from $B(E2)$ values measured by Coulomb excitation and/or lifetime measurements, cf. equations (3.8) and (3.17). Note the Q_0 values, deduced from the ground-state spectroscopic quadrupole moments, viz. in (a) ^{175}Lu, $Q_s = +3.49(2)$ and in (b) ^{159}Tb, $Q_s = +1.432(8)$, which follow from equation (3.19). The Nilsson configurations are (a) $7/2^+[404\downarrow]$ (^{175}Lu), (b) $1/2^+[411\downarrow]$ (^{169}Tm), and (c) $3/2^+[411\uparrow]$ (^{159}Tb). Extended video-based exercises related to ^{175}Lu and intrinsic quadrupole moments are found in exercise 3-6.

Note that as the spin increases, viz. $I = K, K + 1, K + 2, \ldots, Q(I, K)$ may change sign. Also, observe that for $K \rightarrow$ very large, $Q(K, K) \rightarrow Q_0$. Thus, as should be expected, such a system would appear 'classical': this can be seen from figure 3.2 for $m = +l$, $l \rightarrow$ very large. The quantities $Q(I, K)$ are called spectroscopic quadrupole moments, Q_s (we underline, these are the moments that are measured). Some Q_0 values deduced from measured Q_s values are presented in figure 3.14(a)–(c).

3.3 Even-mass nuclei, intrinsic excitations, and rotations

The extent to which the concepts of intrinsic excitation and rotations can be used to organize excited-state data for a deformed nucleus is illustrated in figure 3.16. Figure 3.16 shows 70 excited states in ^{168}Er, assigned uniquely to 14 rotational bands: this constitutes a powerful organizing principle for an enormous body of

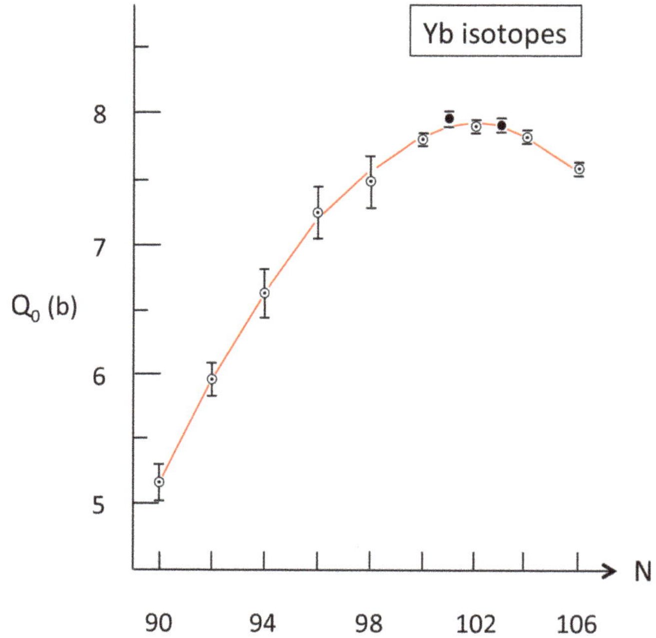

Figure 3.15. Intrinsic quadrupole moments, Q_0 for the even–even and odd-mass ytterbium isotopes. The systematic trend suggests that nuclear deformation may change in a smooth manner along an isotopic sequence.

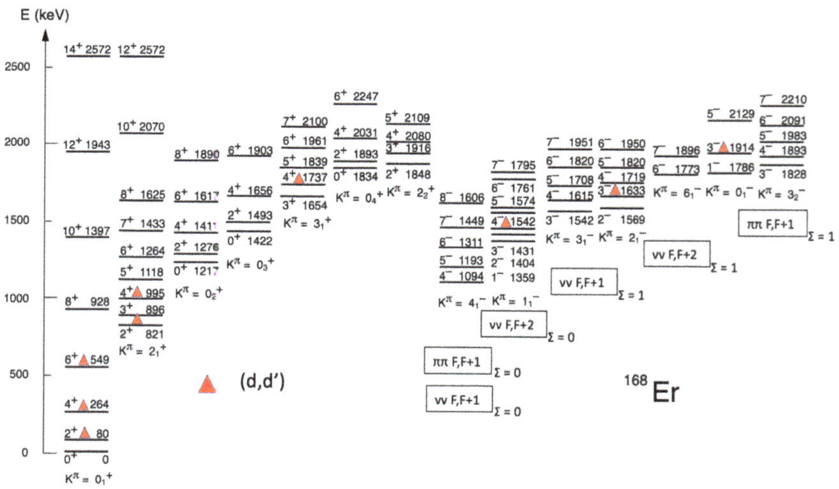

Figure 3.16. Excited states in the even–even nucleus, ^{168}Er organized into rotational bands. This illustrates the power of the rotational model to organize large numbers of excited states into a much smaller number of band structures. Details are discussed in the text. The states populated by the inelastic scattering of deuterons, (d,d′), cf. figure 3.17, are indicated with red triangles. The figure is adapted from [4], copyright (2010) World Scientific Publishing Company.

data. The task at hand is to provide an introductory guide to the interpretation of the various bands shown in figure 3.16, i.e. the intrinsic excitations.

Figure 3.17 shows the population of excited states in ^{168}Er by inelastic deuteron scattering. Besides the first few members of the ground-state band, only a few excited states are populated. In particular, the first intrinsic excitation, the head of a band with $K^{\pi} = 2^{+}$, is populated (also the 4^{+} member of this band). Deformed nuclei widely exhibit first intrinsic excitations with $K^{\pi} = 2^{+}$. Second only to ground-state bands, these $K^{\pi} = 2^{+}$ bands are strongly populated in Coulomb excitation: this is shown for ^{166}Er, a neighbor of ^{168}Er, in figure 3.18. Remarkably, the only consensus regarding the nature of these $K^{\pi} = 2^{+}$ bands is that they involve a quadrupole collective degree of freedom; but whether the intrinsic excitation is rotational or vibrational is unclear (see chapter 5). For present purposes, we refer to such bands as 'collective' and use the popular name of 'gamma' bands when referring to them, cf. figure 5.20.

A further observation with respect to figure 3.17 is that inelastic deuteron scattering also populates a number of states in ^{168}Er with spin-parity 3^{-}. This is a widely occurring feature observed via inelastic scattering in strongly deformed

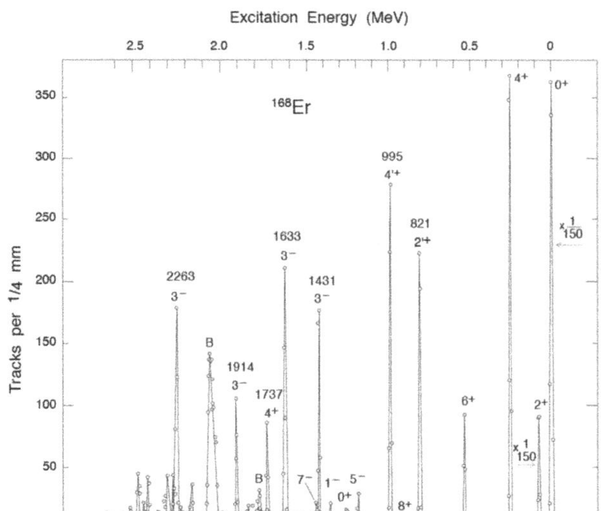

Figure 3.17. Population of excited states in ^{168}Er by inelastic scattering of deuterons. The data are for an incident energy of 12.098 MeV and a scattering angle of 125° with respect to the incident beam. Events are counts per 1/4 mm interval as a function of distance in cm along a photographic plate in the focal plane of a magnetic spectrometer, cf. chapter 6, figure 6.13. Deuteron lines are labeled by the spins and parities of the corresponding levels (cf. figure 3.16). Lines marked with a B are background events resulting from target contaminants. Note the scale reduction factor for the ground state and first excited 2^{+} state. See the text for a discussion of the ^{168}Er levels observed. The strongly populated states are indicated in figure 3.16 by red triangles. The spin-parities and energies of states are from Nuclear Data Sheets. The figure is reproduced from [4], copyright (2010) World Scientific Publishing Company. The original data from which this figure are derived appears in [7].

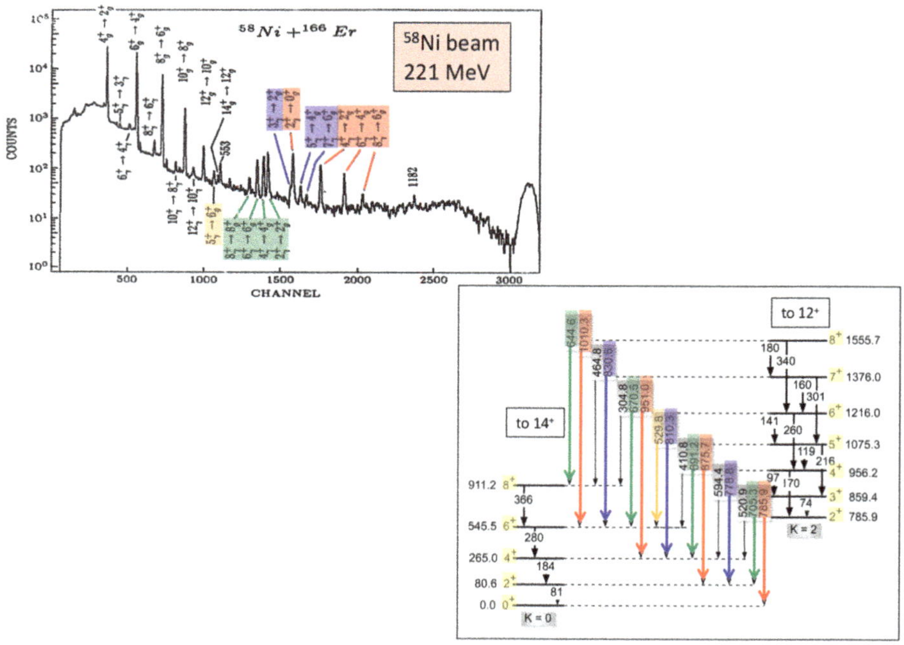

Figure 3.18. Population of excited states in ^{166}Er by Coulomb excitation. Besides the ground-state band, there is significant population of the first excited band which has $K = 2$. Details are discussed in the text. The decay scheme is adapted from [8]. (Reprinted from [9], copyright (1992) with the permission of Elsevier.)

nuclei, with variable numbers of such negative-parity states observed. They are interpreted as resulting from intrinsic collectivity and are termed collective octupole bands. In ^{168}Er, inspection of the bands shown in figure 3.16 reveals that of the four 3^- states, three are members of bands with band heads $K^\pi(E_x) = 1^-(1359)$, $2^-(1569)$ and $0^-(1786)$; the fourth 3^- state at 2263 keV is assigned to a $K^\pi = 3^-$ band with this state as the band head. However, we note that there are other bands ($K^\pi = 3^-$ at 1542 and 1828 keV) that possess 3^- states and these states are not populated by inelastic scattering of deuterons. There are isolated mass regions where negative-parity bands occur at very low energy. The lowest known occurrence is for $A \sim 224$ and this is illustrated in figure 3.19.

Many other bands, involving what must be intrinsic excitation, are evident in figure 3.16. These bands result from 'breaking pairs', i.e. recoupling $J = 0$ pairs to resultant spin with $J \neq 0$. These broken pairs are Nilsson states and so the value of J is dictated by the Nilsson quantum number, Ω as $\Omega_1 + \Omega_2$ or $|\Omega_1 - \Omega_2|$. This occurs in odd–odd deformed nuclei already at the ground state, as shown in figure 3.20. In even–even nuclei, the breaking of a $J = 0$ pair requires an energy of ~1.5 MeV because of the pairing correlations. A discussion of these issues is beyond the present level of detail; but will receive detailed discussion later in the Series. (The labels $\Sigma = 0$ and $\Sigma = 1$ indicate intrinsic spin couplings with the Nilsson configurations: they are not important for the present discussion.)

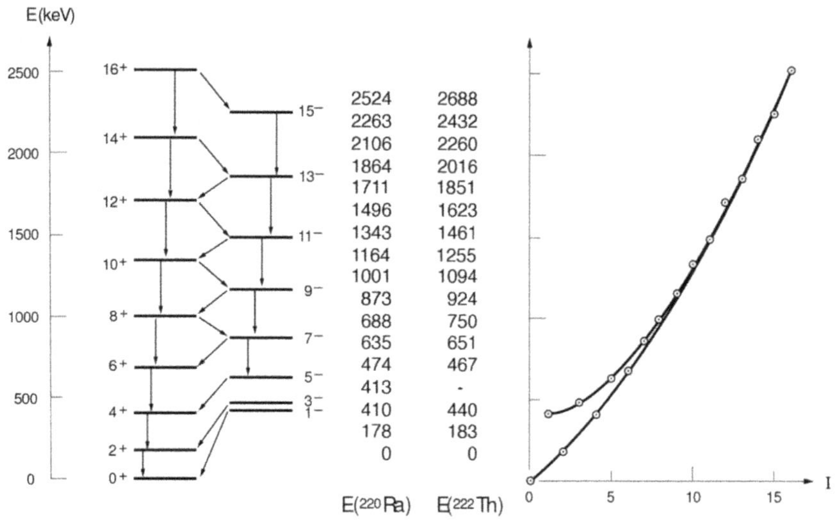

Figure 3.19. Low-lying negative parity rotational states in ^{220}Ra and ^{222}Th (the 1^- state has not yet been identified in ^{222}Th). The plot on the right gives a perspective on the common features of the positive- and negative-parity bands in these nuclei. The consensus view of these structures is that they involve octupole deformation, but what underlies the separate low-spin energy patterns does not have a clear explanation. The figure is reproduced from [4], copyright (2010) World Scientific Publishing Company. (The data are taken from Nuclear Data Sheets.)

3.4 Deformed nuclei and the *K* quantum number

Deformed nuclei exhibit some widely occurring features that are distinctive, and can be understood using the ideas that have now been developed. These features are treated in subsections. They are associated with the *K* quantum number.

3.4.1 *K* isomerism

Deformed nuclei sometimes exhibit excited rotational bands that have band heads that are long lived. They are characterized by a high value for the *K* quantum number. An exceptional example is the nucleus ^{178}Hf, details of which are shown in figure 3.21. (Indeed, this nucleus and the $K^\pi = 8^-$ state are featured in figure 1.5.) These band heads are called *K* isomers. A summary of the longest-lived cases is presented in table 3.1.

The reason for the long half-lives of such states is due to the large change in the *K* quantum number when the spin change is small. Electromagnetic transitions in nuclei are dominated by transitions with low multipolarity. This is understood at the level of the Weisskopf estimates for electromagnetic transitions, shown in table 3.1. For example, the decay of the 8^- state in ^{178}Hf at 1147 keV could, in principle, occur by any transition to a lower state, e.g. by a $\Delta I = 8$, $\Delta\pi$ = yes transition to the ground state. The Weisskopf estimate for such a transition gives a half-life of 5×10^{10} yr. The 'compromise' is the decay path shown via an *E*1 transition, which would

Figure 3.20. Excited states in the odd–odd nucleus, ^{166}Ho organized into rotational bands. The Nilsson configurations of the bands are deduced from the possible Nilsson states expected for $Z = 67$ and $N = 99$ with combinations p $\Omega^{\pi}[Nn_z\Lambda]\pm$ n $\Omega^{\pi}[Nn_z\Lambda]$. The relative energy of these two couplings is beyond the present level of discussion.

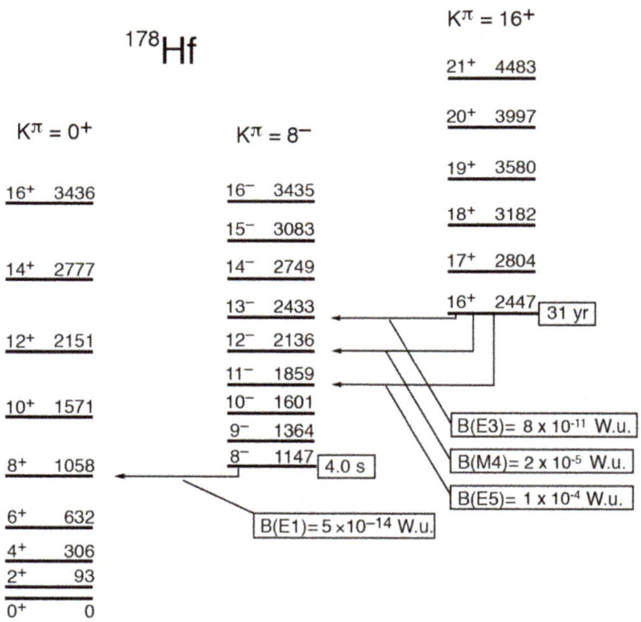

Figure 3.21. The isotope ^{178}Hf exhibits a variety of K isomers. These occur as the heads of high-K rotational bands and the half lives are given. The decay branches with their strengths in W.u. are shown. The figure is reproduced from [4], copyright (2010) World Scientific Publishing Company.

Table 3.1. Selected K isomer decays.

Isotope	$M\lambda$ or $E\lambda$	E_i (keV)	E_γ (keV)	$I_i, K_i^{\pi_i} \rightarrow I_f, K_f^{\pi_f}$	ΔK	Trans. rate (W.u.)
^{244}Cm	$M1$	1040	744	$6, 6^+ \rightarrow 6, 0^+$	6	5.6×10^{-13}
^{174}Yb		1518	992	$6, 6^+ \rightarrow 6, 0^+$	6	7.0×10^{-12}
^{178}Hf		2574	141	$14, 14^- \rightarrow 13, 8^-$	6	1.8×10^{-8}
^{174}Yb	$E2$	1518	992	$6, 6^+ \rightarrow 6, 0^+$	6	8.5×10^{-9}
^{244}Cm		1040	744	$6, 6^+ \rightarrow 4, 0^+$	6	2.4×10^{-10}
^{174}Yb		1518	1265	$6, 6^+ \rightarrow 4, 0^+$	6	8.7×10^{-11}
^{178}Hf	$E1$	1147	89	$8, 8^- \rightarrow 8, 0^+$	8	5.1×10^{-14}
^{180}Hf		1141	58	$8, 8^- \rightarrow 8, 0^+$	8	3.5×10^{-17}
^{182}Hf		1173	51	$8, 8^- \rightarrow 8, 0^+$	8	1.3×10^{-16}
^{184}Hf		1272	73	$8, 8^- \rightarrow 8, 0^+$	8	3.5×10^{-15}
^{180}Hf	$M2$	1141	501	$8, 8^- \rightarrow 6, 0^+$	8	9.8×10^{-15}
^{182}Hf		1173	507	$8, 8^- \rightarrow 6, 0^+$	8	4.0×10^{-12}
^{178}Hf	$E3$	2447	13	$16, 16^+ \rightarrow 13, 8^-$	8	8.1×10^{-11}
^{180}Hf		1141	501	$8, 8^- \rightarrow 6, 0^+$	8	7.1×10^{-10}
^{178}Hf	$M4$	2447	310	$16, 16^+ \rightarrow 12, 8^-$	8	2.4×10^{-5}
^{178}Hf	$E5$	2447	587	$16, 16^+ \rightarrow 11, 8^-$	8	1.2×10^{-4}

normally be fast. However, a spin change of 1 cannot effect a change in K quantum number of 8.

One nucleus where such K hindrance plays a major role is ^{180}Ta where the ground state ($I^\pi = 1^+$) has a half-life of 8.154 h; and above a rotational excited state ($I^\pi = 2^+$, 39.54 keV) is a state at 77.2 keV with a half-life of $> 7.1 \times 10^{15}$ y. This state has $I^\pi = 9^-$ and $K = 9$. It could in principle decay by an $E7$ electromagnetic transition of 37.7 keV. The drama is that the isotope ^{180}Ta should not be found in Nature; but it *is*, due entirely to K isomerism.

The presence of K isomers in deformed nuclei can lead to competition between β decay and γ decay. An example is shown in figure 3.22. Normally, γ decay is much faster than β decay; but when γ decay is hindered by a large change in the K quantum number, β decay may be observed in competition with γ decay (by way of decaying to a daughter nucleus with K_f value = K_i, $K_i \pm 1$).

3.4.2 One-nucleon transfer reactions

One-nucleon transfer reactions are a powerful spectroscopic probe of states in nuclei that have some degree of independent particle (or hole) character. Figure 3.23(a)–(c) illustrates the spectroscopic selectivity of the one-proton pickup reaction, (t,α) leading to excited states in ^{165,167,169}Ho. The population of the different spin states within each band, defined by the quantum number K, has a characteristic pattern: it is called a transfer reaction 'spectroscopic fingerprint'. The explanation comes from the distinction between the laboratory frame, wherein the transfer of the nucleon takes place, and the body frame, wherein the quantum number K is defined. By expanding the laboratory frame state in terms of the body frame states, viz.

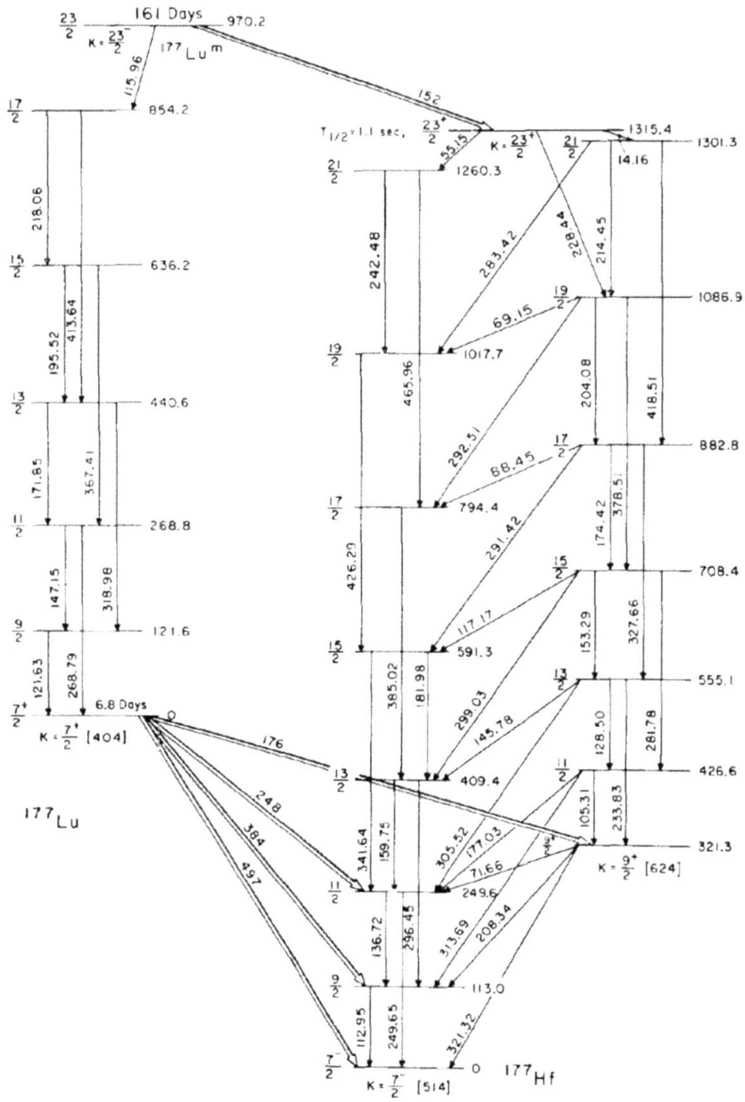

Figure 3.22. Beta decay of a K isomer in ^{177}Lu to similar states in ^{177}Hf. The heavy arrows indicate the β-decay branches deduced from the observed γ-ray transitions. Reprinted from [10], copyright (1967) with the permission of Elsevier.

$$|IMK\rangle = \sum_{jl} C_{jl}|jlMK\rangle, \tag{3.20}$$

where $K = m_j$, the spectroscopic fingerprints can be calculated.

3.4.3 Band mixing and crossing

Bands in deformed nuclei can undergo quantum mechanical state mixing. This can result in dramatic effects when bands intersect or 'cross'. We emphasize here the

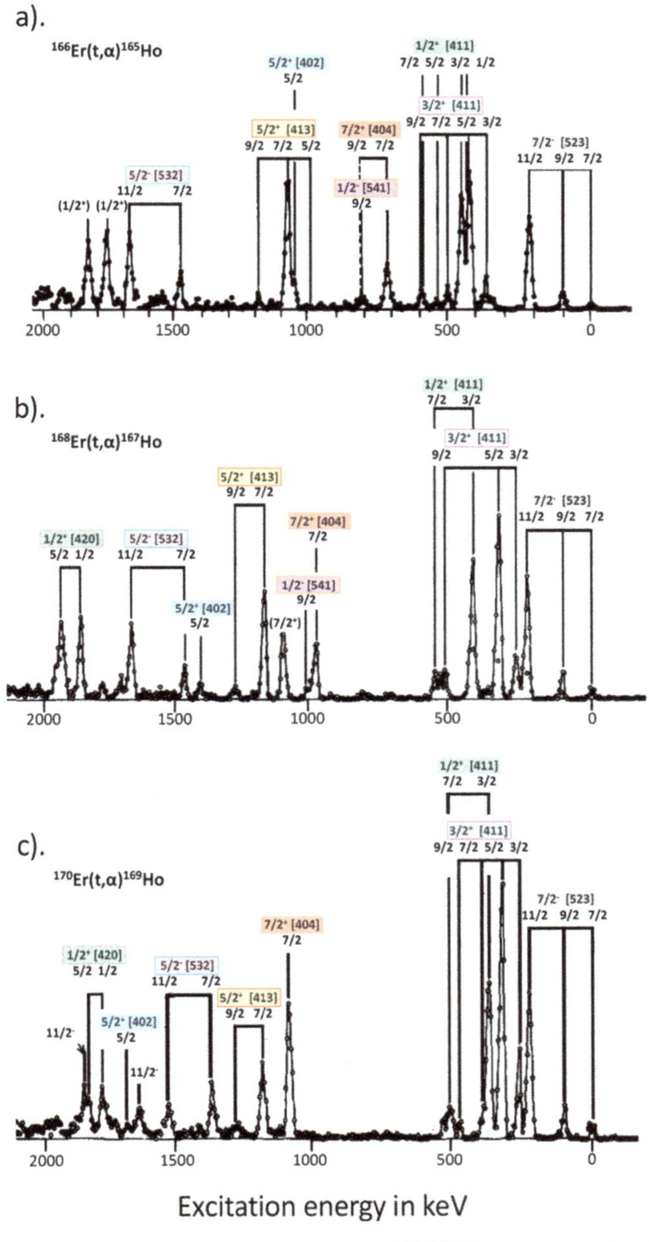

Figure 3.23. (a)–(c) One-proton pickup reactions leading to ^{165,167,169}Ho. The rotational bands, which are characteristic of Nilsson configurations near $Z = 67$, are shown. Note the selectivity of this process for population of the various band members. See the text for some more details. The figure is adapted from figures that appear in [11], copyright (1978), and [12], copyright (1979), with permission from Elsevier.

basic feature of quantum mechanical state mixing that only states with the same spin and parity can mix in a finite many-body quantum system.

When bands 'cross', if they have the same K quantum number and the same parity, they will undergo quantum mechanical state mixing. An example is shown in

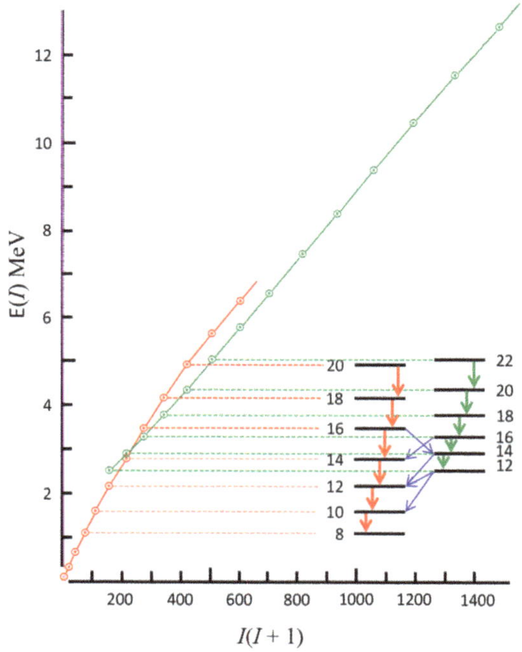

Figure 3.24. The ground-state band (red points) and the excited band (green points) which becomes the yrast band at spin 16 in ^{166}Yb.

figures 3.24 and 3.25(a) and (b). Such behavior is widespread in deformed nuclei and has often been described by the term 'backbending'. (The mixing of bands with different K quantum numbers is handled in chapter 5.)

3.4.4 Magnetic moments

Nuclear magnetic moments are non-zero whenever the total nuclear spin is not zero. A magnetic moment is related to the nuclear spin as

$$\boldsymbol{\mu} = g\mathbf{I}, \tag{3.21}$$

where g is called the 'g factor'. The simplest global view of nuclei is for $g(2^+_1)$ values and this is shown for deformed nuclei in the rare-earth region in figure 3.26. The ratio Z/A is the simplest model estimate for $g(2^+_1)$: evidently this is approximate and there is a lack of any simple systematic trend. Rotational bands are characterized by two contributions to g, viz.

$$g = g_R + (g_K - g_R)\frac{K^2}{I(I+1)}, \tag{3.22}$$

where g_K is a characteristic of the intrinsic configuration of a band with K quantum number and g_R is the collective rotational contribution which is due to the rotating electrical charge of the nucleus and varies with R.

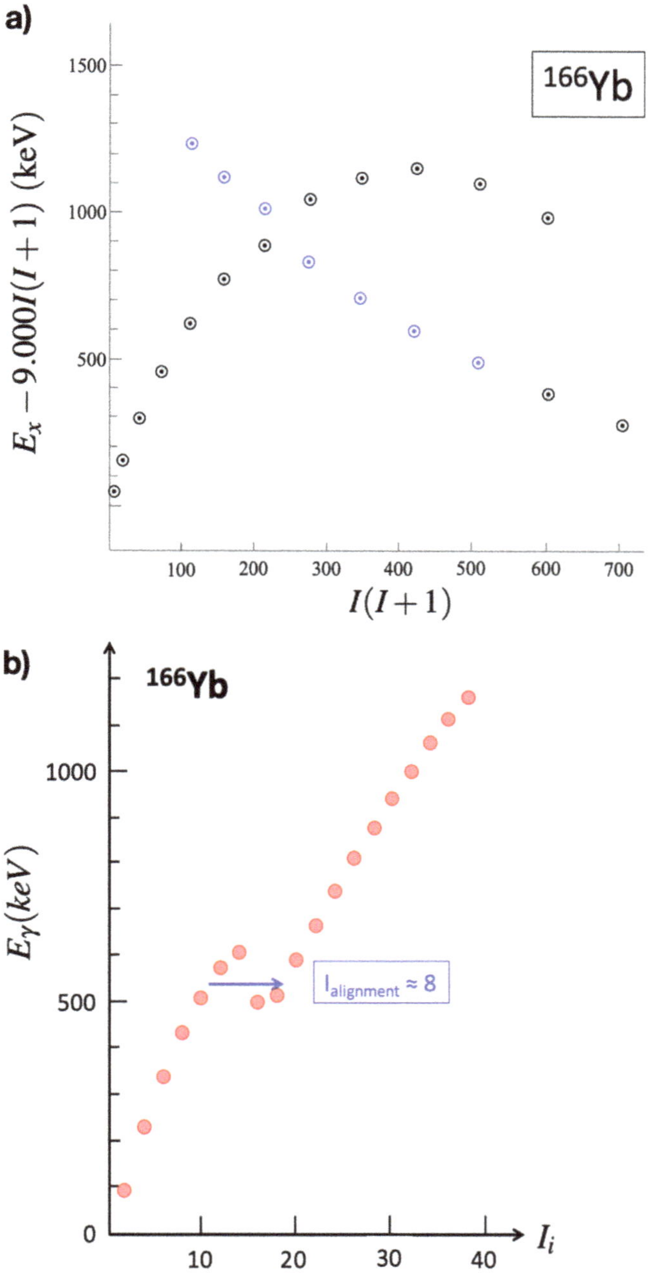

Figure 3.25. (a) and (b) Band crossing and mixing in ^{166}Yb. Note the energy scale in (a) is unusual in that a 'rotational reference energy' of $9.000I(I+1)$ keV has been subtracted from the experimental excitation energies: this is simply to exaggerate the systematic features of interest so that they can be seen easily. These are plotted against $I(I+1)$. In (b) γ-ray transition energies are plotted versus spin of the initial state for the transition for the yrast states. This is similar to the plots in figure 3.9(a)–(d), and suggest that the band above the crossing, cf. (a) has an 'alignment' of $8\hbar$.

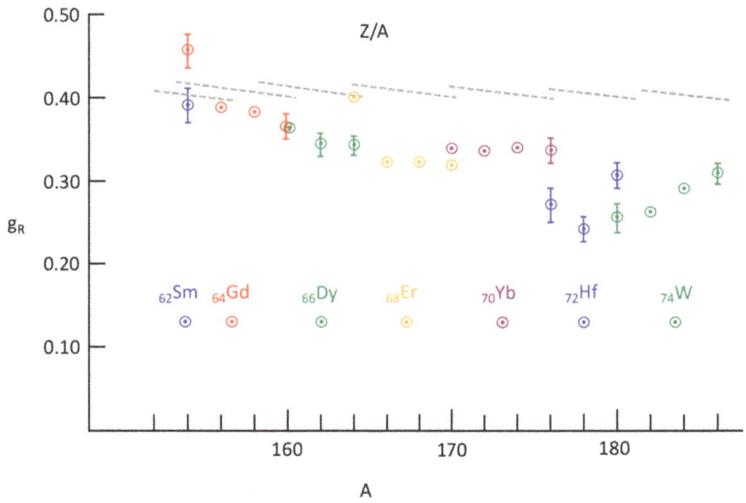

Figure 3.26. Rotational *g* factors, g_R deduced from magnetic moments for 2_1^+ states in deformed nuclei in the rare-earth region. The estimates given by *Z*/*A* are shown as dashed lines for the *Z* values of the nuclei presented. Data points without error bars possess uncertainties that are less than the sizes of the circles; the selection omits values with low precision. See the text for further details. The magnetic moment values are taken from the ENSDF files.

Atomic hyperfine structure provides information on *g* factors of nuclear ground states and long-lived isomers, cf. figure 1.5. Excited state *g* factors are measured by observing the angular distribution of γ-rays emitted from nuclei produced in a nuclear reaction and recoiling into a magnetic medium: the nuclear magnetic moment precesses and the angular distribution of the γ-rays is modulated. Further details of magnetic properties of nuclei and their measurement will be handled later in the Series.

3.5 Nuclear moments of inertia

Nuclear rotation implies that the concept of a moment of inertia can be defined. This emerges directly from equation (3.2) in the form of the parameter $\mathcal{J}$. This is shown for all rotational nuclei in figure 3.27 where a comparison of $\mathcal{J}_{\text{expt}}/\mathcal{J}_{\text{rigid}}$ versus β, the quadrupole deformation parameter, is presented. The quantities plotted are obtained using the relationships, which have dimensions of $\text{m}^2 \cdot\text{kg}$,

$$\mathcal{J}_{\text{expt}} = 0.208\,0 \times 10^{-54} E(2_1^+)^{-1}[\text{MeV}^{-1}], \tag{3.23}$$

$$\mathcal{J}_{\text{rigid}} = 0.886\,4 \times 10^{-57} A^{5/3}(1 + 0.315\,4\beta + 0.44\beta^2), \tag{3.24}$$

$$\mathcal{J}_{\text{irrot}} = 0.886\,4 \times 10^{-57} A^{5/3} 0.895\,1\beta^2, \tag{3.25}$$

where

$$\beta = Q_0 \frac{\sqrt{5\pi}}{3ZR^2}, \tag{3.26}$$

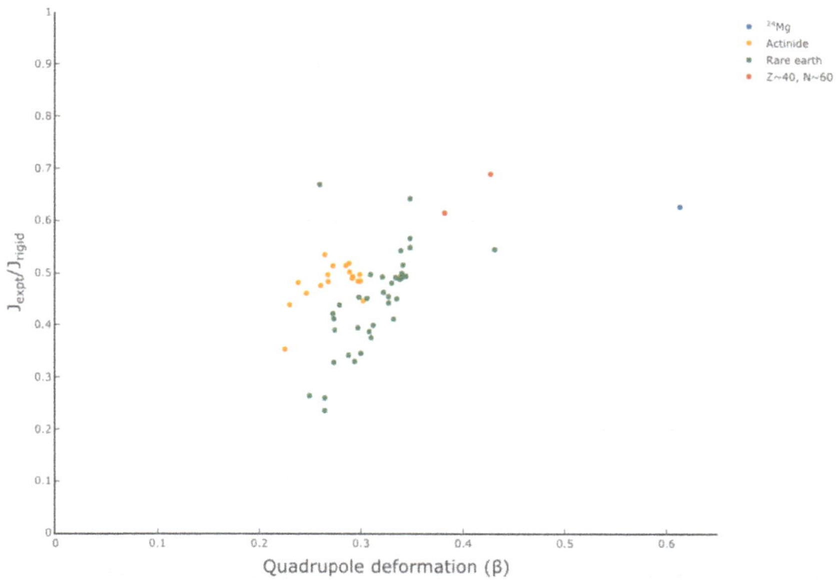

Figure 3.27. Selected view of nuclear moments of inertia for nuclei which exhibit a rotational band built on their ground state. Moments of inertia are presented as a ratio of the experimental moment of inertia deduced from the excitation energy of the 2_1^+ state (equation (3.23)) divided by the rigid body moment of inertia (equation (3.24)), plotted as a function of the quadrupole deformation, β, for the relevant isotopes. Data are presented for the rare earth isotopes, the actinide region, the region close to $Z \sim 40$, $N \sim 60$ and for ^{24}Mg. Animation available at https://iopscience.iop.org/book/978-0-7503-2674-2.

with Q_0 obtained from $B(E2)$ values, cf. equations (3.6) and (3.8), and $R = 1.2A^{1/3}$ fm. The relationships, equations (3.23)–(3.26), are discussed in chapter 5.

Naively, one expects the moment of inertia to be precisely defined by the mass distribution of the rotating body. There are many theories that attempt to describe nuclear moments of inertia, but none of them has gained wide acceptance as the correct view.

The physics of nuclear rotations, which are quantum mechanical, are in contrast to another system that exhibits quantum mechanical rotations, namely molecules. Comparison between nuclear and molecular rotation is made in chapter 5.

Nuclear moments of inertia as a function of increasing spin in individual nuclei can be extracted using equation (3.2). Some results are shown in figure 3.28. Evidently, curvature in the plots reveals that what is being termed a moment of inertia may increase with spin. This may seem reasonable on the basis of the liquid-like drop property of the nucleus, i.e a rotating liquid drop should undergo centrifugal stretching with increasing spin. But figure 3.5 contradicts this: the intrinsic quadrupole parameter, Q_0 extracted from $B(E2)$ values and $E2$ matrix elements, for the same nuclei, indicate that there is no evidence for an increase in the quadrupole parameter. Indeed, there are examples of so-called superdeformed bands in nuclei which show energy spacing consistent with a near rigid rotational energy parameter, as depicted in figure 3.29. Superdeformed bands are discussed further in chapter 4.

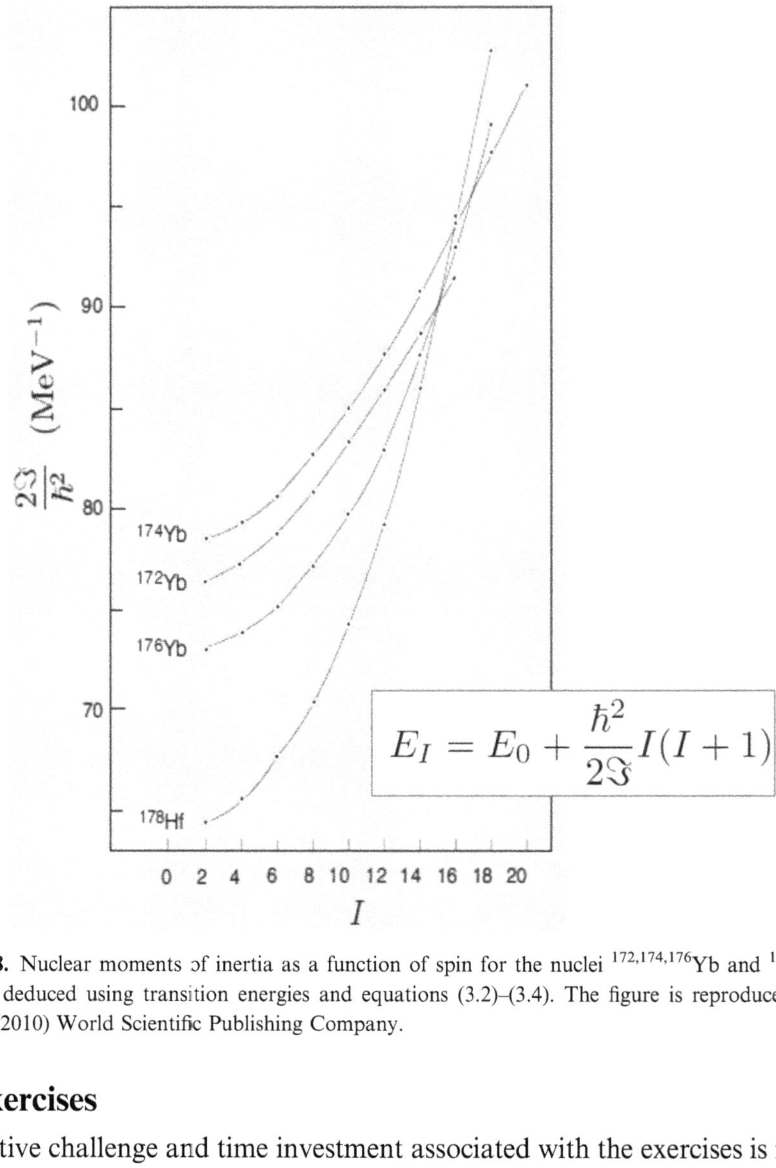

Figure 3.28. Nuclear moments of inertia as a function of spin for the nuclei ^{172,174,176}Yb and ^{178}Hf. These values are deduced using transition energies and equations (3.2)–(3.4). The figure is reproduced from [4], copyright (2010) World Scientific Publishing Company.

3.6 Exercises

The relative challenge and time investment associated with the exercises is indicated from * to ***. Exercises marked with '**E**' require interactive access to figures or watching included video exercises and will only have such functionality in the e-book version.

3-1. Figure 3.4 presents the ground state rotatational bands in ^{242}Pu and ^{244}Pu along with the excitation energies of the rotational states. Scale the excitation energies to that of the 2_1^+ states, i.e. divide the energies for ^{242}Pu by 45 and the energies for ^{244}Pu by 46. Having scaled the energies, plot the two bands as $E(I;$ scaled) versus $I(I + 1)$. Note that, while the energies look significantly different, when scaled they are nearly identical, i.e. there is evidently some underlying 'universal' rotational energy profile.

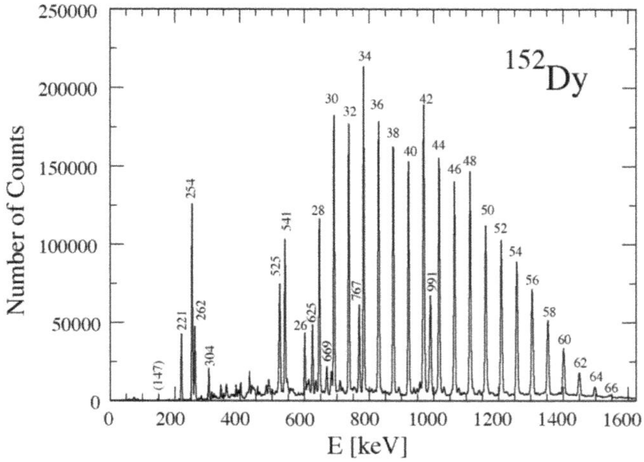

Figure 3.29. A γ-ray spectrum for a superdeformed band in ^{152}Dy (see section 4.1). Note the near-constant spacing of γ-ray energies and the connection to $\Delta^2 E$, cf. rotor model and equation (3.3). (The decrease in γ-ray intensities below spin 30 indicates the 'draining' out of the band, which is complete by spin 26.) The figure is reproduced from [4], copyright (2010) World Scientific Publishing Company. The original data are presented in [13].

This does not appear to have been recognized; but holds for the most strongly deformed ground-state bands, extending into the rare earth region, e.g. ^{174}Yb. (Higher precision energy values can be used; but this exercise captures the feature.) **

3-2. Figure 3.8(a) shows a partial level scheme for ^{175}Lu which includes two $K = 1/2$ rotational bands: $1/2^-$[541] and $1/2^+$[411] labeled in violet and green, respectively.
   (a) Make a comparative plot of E versus $I(I + 1)$ for each of the ^{175}Lu bands in figure 3.8(a). *
   (b) Make use of figure 3.12 to suggest values for the decoupling parameter, a, which would be the most consistent with the pattern of excited states in each of the two $K = 1/2$ bands. *
   (c) Using the estimate for the decoupling parameter, a, make use of equation (3.16) to estimate the value of the scale factor, A, for both $K = 1/2$ bands. *
   (d) Having obtained values for both a and A, test your function for $E(I)$ with respect to your initial plot in part (a). *

3-3. Using figure 3.12, predict the ground state spin/parity of deformed nuclei whose ground state band has $K = 1/2$ when the decoupling parameter, $a = -8, -3, +3$, and $+6$. *

3-4. Figure 3.27 presents a comparison of experimental moments of inertia (obtained using equation (3.23)) to the rigid body moment of inertia (calculated with equation (3.24)). Use the NUDAT cross-variable plot to locate $E(2_1^+)$ and β values for a selection of the nuclei shown on figure 3.27 (the interactive figure in the e-book version will allow you to see which isotopes are plotted). Make your own conclusions on whether irrotational flow (equation (3.25)) is a better correspondence to experimental data than rigid body by making a plot of $\mathcal{J}_{\text{expt}}/\mathcal{J}_{\text{irrot}}$ versus β. *** **E**

3-5. Following the procedures introduced for extracting $B(E2)$ and $\rho^2(E0)$ values in ^{152}Sm, and consulting the ENSDF pages for ^{154}Gd (which shows a nearly identical pattern of excited states and decays as ^{152}Sm), obtain $B(E2)$ and $\rho^2(E0)$ values for the corresponding transitions in ^{154}Gd, i.e. for the various decays of the 2_1^+, 0_2^+, and 2_2^+ states. *** **E**

3-6. For those reading the e-book, explore the set of video-based exercises presented in figure 3.30–3.35. These videos are short self-contained explanations of different aspects of the nuclear structure of ^{175}Lu and contain short exercises intended to encourage familiarity with the Nilsson model and to draw on many of the concepts and equations presented in this chapter. The powerpoints presented in the short videos can also be accessed here: [https://iopscience.iop.org/book/978-0-7503-2674-2]. *** **E**

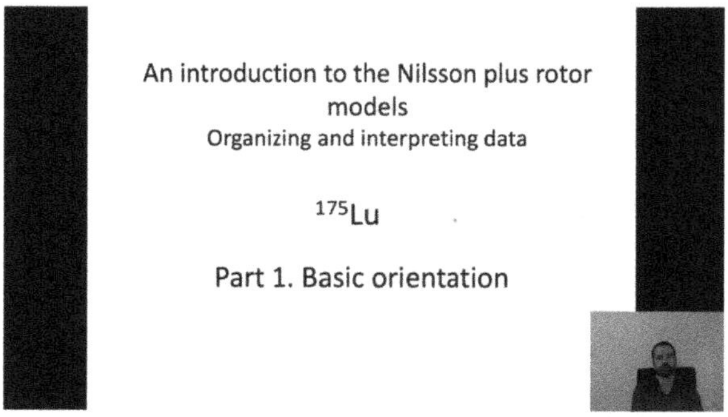

Figure 3.30. Video exercise on ^{175}Lu: Part 1. Basic orientation. Video available at https://iopscience.iop.org/book/978-0-7503-2674-2.

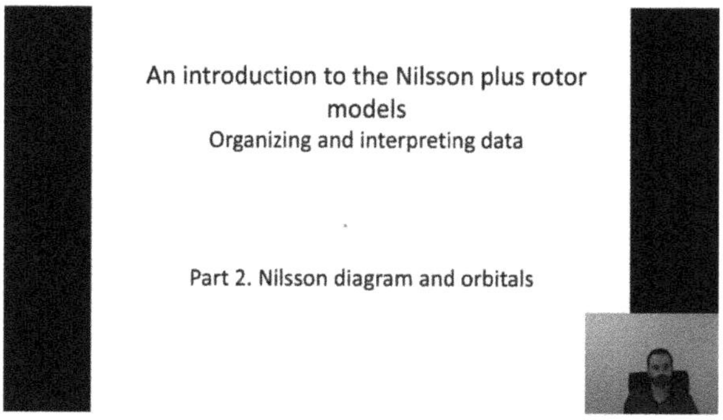

Figure 3.31. Video exercise on ^{175}Lu: Part 2. Nilsson diagram and orbitals. Video available at https://iopscience.iop.org/book/978-0-7503-2674-2.

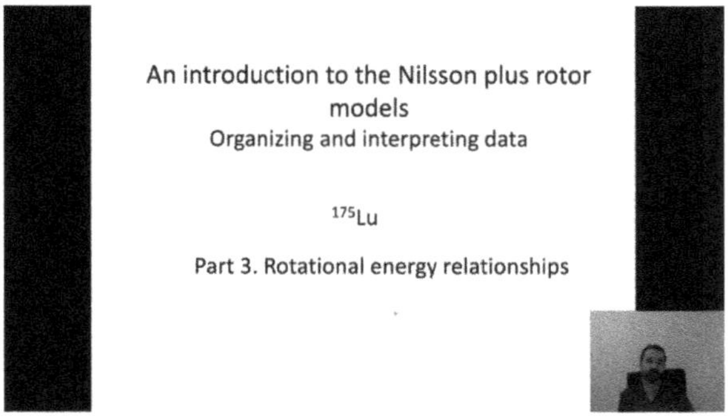

Figure 3.32. Video exercise on ^{175}Lu: Part 3. Rotational energy relationships. Video available at https://iopscience.iop.org/book/978-0-7503-2674-2.

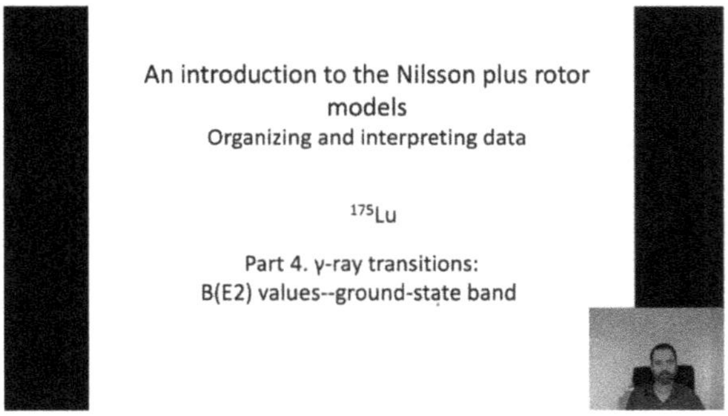

Figure 3.33. Video exercise on ^{175}Lu: Part 4. γ-ray transitions. Video available at https://iopscience.iop.org/book/978-0-7503-2674-2.

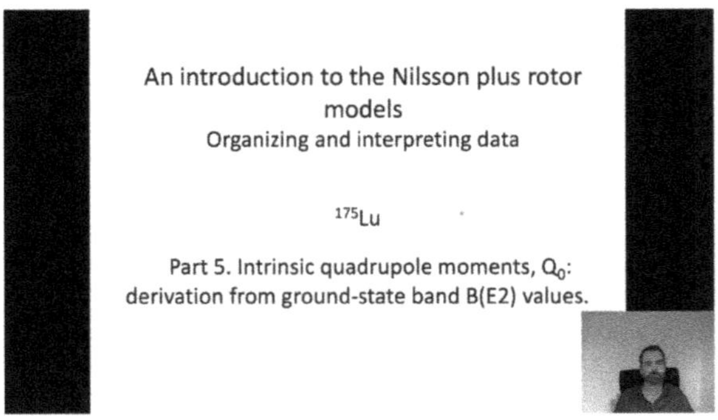

Figure 3.34. Video exercise on ^{175}Lu: Part 5. Intrinisic quadrupole moments—derivation from ground-state band B(E2) values. Video available at https://iopscience.iop.org/book/978-0-7503-2674-2.

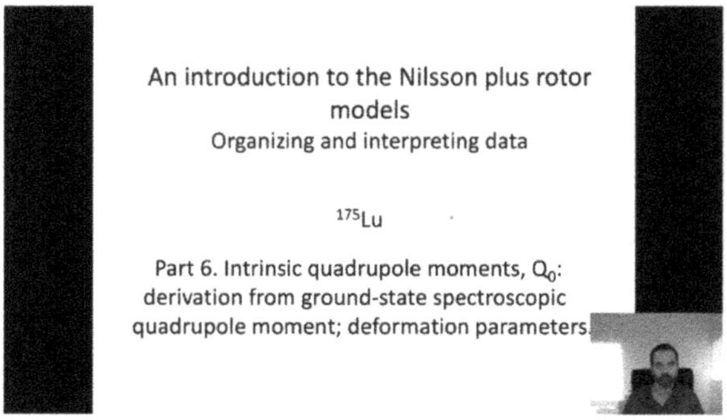

Figure 3.35. Video exercise on ^{175}Lu: Part 6. Intrinisic quadrupole moments—derivation from ground-state spectroscopic quadrupole moment. Video available at https://iopscience.iop.org/book/978-0-7503-2674-2.

References

[1] Nakatsukasa T, Matsuyanagi K, Matsuzaki M and Shimizu Y R 2016 Quantal rotation and its coupling to intrinsic motion in nuclei *Phys. Scr.* **91** 073008

[2] Nilsson S G 1955 Binding states of individual nucleons in strongly deformed nuclei *Mat. Fys. Medd. Dan. Vid. Selsk.* **29** 1

[3] Chupp T E, Fierlinger P, Ramsey-Musolf M J and Singh J T 2019 Electric dipole moments of atoms, molecules, nuclei and particles *Rev. Mod. Phys.* **91** 015001

[4] Rowe D J and Wood J L 2010 *Fundamentals of Nuclear Models: Foundational Models* (Singapore: World Scientific)

[5] Spreng W *et al.* 1983 First observation of backbending in an actinide nucleus *Phys. Rev. Lett.* **51** 1522

[6] Skensved P, Chapman R, Kozub R L, Leslie J R, McLatchie W, Ward D, Andrews H R and Häusser O 1981 The ground-state rotational band in ^{175}Lu *Nucl. Phys.* A **366** 125

[7] Tjøm P O and Elbek B 1968 Collective vibrational states in even Erbium nuclei *Nucl. Phys.* A **107** 385

[8] Kulp W D, Allmond J M, Hatcher P, Wood J L, Loats J, Schmelzenbach P, Stapels C J, Krane K S, Larimer R-M and Norman E B 2006 Precision test of the rotor model from band mixing in ^{166}Er *Phys. Rev.* C **73** 014308

[9] Fahlander C *et al.* 1992 Triaxiality in ^{166}Er *Nucl. Phys.* A **537** 183

[10] Bernthal F M and Rasmussen J O 1967 Influence of coriolis coupling, pairing and octupole vibration–particle coupling on $\Delta K = -1$ $E1$ transitions in ^{177}Hf *Nucl. Phys.* A **101** 513

[11] Løvhøiden G, Burke D G, Hirning C R, Flynn E R and Sunier J W 1978 Levels in ^{165}Ho populated in the ^{166}Er(t,α) reaction *Nucl. Phys.* A **303** 1

[12] Løvhøiden G, Burke D G, Flynn E R and Sunier J W 1979 Levels in ^{167,169}Ho populated in the (t,α) reaction *Nucl. Phys.* A **315** 90

[13] Lauritsen T *et al.* 2002 Direct decay from the superdeformed band to the yrast line in ^{152}Dy$_{86}$ *Phys. Rev. Lett.* **88** 042501

Chapter 4

Towards a global view of nuclear structure

A first look at shape coexistence in nuclei, non-deformed (spherical) nuclei, and weakly deformed nuclei is made. The association of E0 transitions with coexisting shapes and their mixing is outlined. Pairing correlations and their dominance of the structure of spherical nuclei is introduced. The concept of a Cooper pair is sketched. Basic details of double-closed shell nuclei are presented. A limited look at excitation patterns observed in weakly deformed nuclei is provided: we underline the incompleteness of this treatment.

Concepts: shape coexistence, mixing, $E0$ transition strength, intruder states, seniority, pairing correlations, Cooper pair, weak coupling, rotation-aligned coupling, triaxiality.

A global view of nuclear structure does not yet exist. The primary reason is that key new discoveries are being made that necessitate major changes in viewpoint. A secondary reason is that the database is very incomplete with respect to nuclei that are neither spherical nor strongly deformed.

Nuclei that can be viewed as spherical are very few in number. Thus, a complete view of the types of structure to be expected in spherical nuclei is lacking. A major change in the interpretation of nuclei viewed as spherical has been the serious doubt placed on the role of low-energy vibrational degrees of freedom in such nuclei. Further, it is particularly in such nuclei that the phenomenon called 'shape coexistence' is widely occurring at low energy. Indeed, it would appear that shape coexistence in nuclei occurs universally. But the manifestation of shape coexistence is often subtle and complex; in consequence, patterns of behavior have been very difficult to elucidate. In this chapter, we introduce structural features that are clearly associated with: shape coexistence in nuclei; spherical (non-deformed) nuclei. We take a limited look at structural features in weakly deformed nuclei.

doi:10.1088/978-0-7503-2674-2ch4

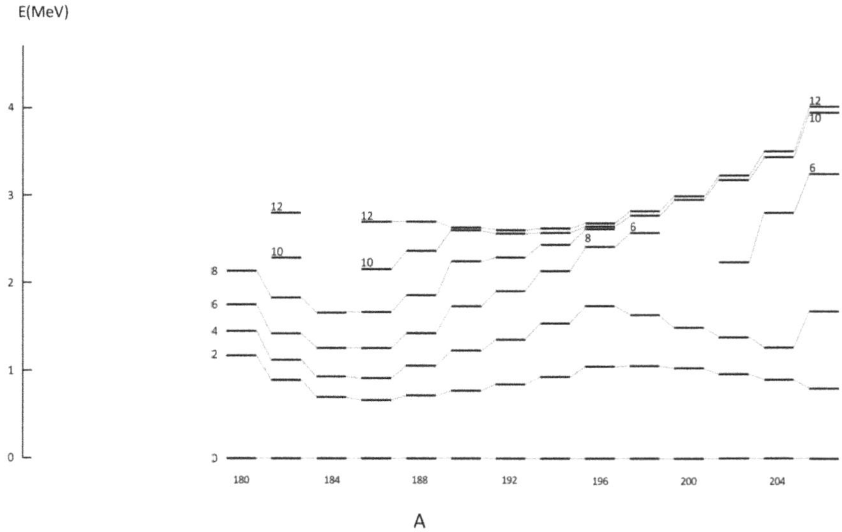

Figure 4.1. Systematics of yrast states in the Pb (Z = 82) isotopes. Details are discussed in the text.

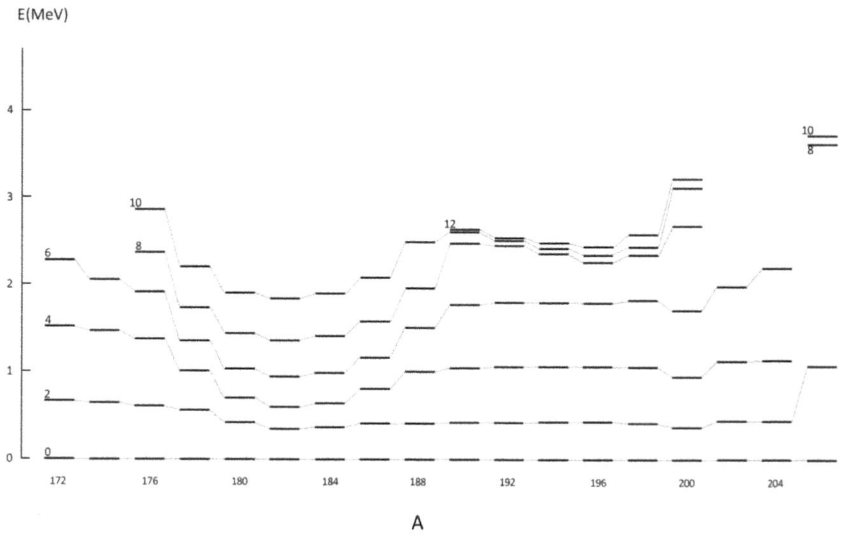

Figure 4.2. Systematics of yrast states in the Hg (Z = 80) isotopes. Details are discussed in the text.

Figure 4.1 shows the so-called 'yrast' states[1] in the even-mass Pb isotopes. They possess a closed proton shell, $Z = 82$. Two features should be noted: in $^{190-206}$Pb, the states with spin 12 and spin 10 always lie very close together; in the mid-shell region (N = 104, i.e. ^{186}Pb) states of relatively high spin occur at relatively low energy. Figure 4.2 shows the yrast states in the even-mass Hg isotopes, which possess $Z = 80$

[1] The word 'yrast' derives from the Swedish word 'yr' (= dizzy); literally the intended meaning is 'dizziest'. Practically, it refers to the highest-spin state at a given excitation energy, e.g. the ground-state rotational band in a deformed nucleus forms a set of yrast states.

and are adjacent to the Pb isotopes. The pattern noted in the even-Pb isotopes persists. Figure 4.3 shows that the patterns occurring in the even-Pb and even-Hg isotopes have similar counterparts in the Sn ($Z = 50$, closed shell) and Cd ($Z = 48$) isotopes. Here, the low-energy, high-spin structures occurring at mid shell, $N = 66$, do not come quite so

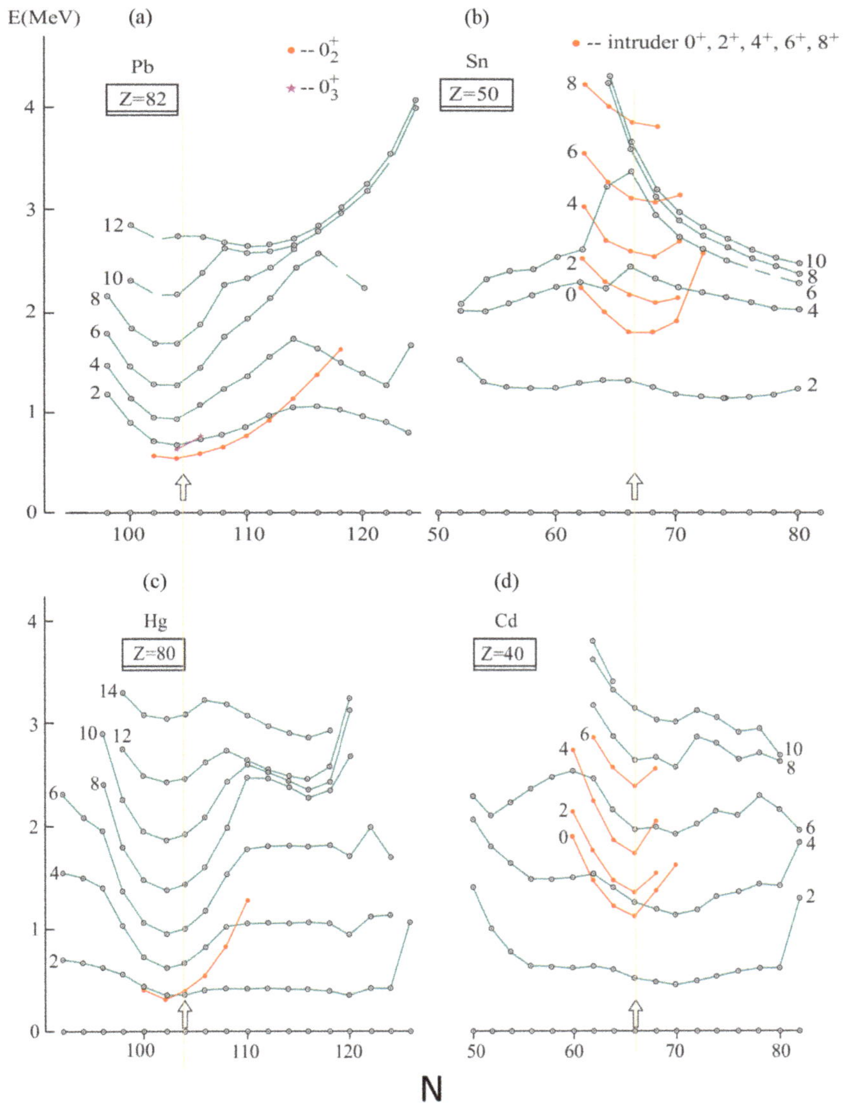

Figure 4.3. Shape coexistence in the $Z = 82$ and $Z = 50$ closed shell nuclei: (a) and (b) the even–even Pb and Sn isotopes; (c) and (d) the even–even Hg and Cd isotopes. The mid-shell points at $N = 104$ and $N = 66$ are indicated. States connected by green lines are yrast states and are easy to populate for experimental study. States connected by red lines are much harder to populate. The second excited 0^+ states in $^{186,188}Pb$ are shown in purple. (The closely spaced high-spin states at high energy for $108 \leqslant N \leqslant 124$ and $66 \leqslant N \leqslant 80$ are so-called 'seniority' structures—see chapter 5.)

low in energy as in the Pb and Hg isotopes and they are not yrast structures. The high-spin states that lie close in energy are characteristic of nuclei dominated by spherical (shell model) configurations with pairing correlations. The high-spin states that appear at lower energies near mid-shell are characteristic of nuclei that exhibit shape coexistence.

In the following discussion, we take a closer look at some examples of shape coexistence in nuclei; followed by a more detailed look at the effect of pairing on shell model configurations. We complete the chapter with the identification of features that appear useful for organizing and, potentially, understanding weakly deformed nuclei.

4.1 Shape coexistence in nuclei

A remarkable property of nuclei is the existence of structures with different deformations. This is a feature beyond the wide range of deformations that occur across nuclear ground states; it involves the 'coexistence' of excited states that can be grouped into structures indicative of more than one deformation in a single nuclear species (A, Z, N). This is termed 'shape coexistence'; also the term 'shape isomerism' is used[2].

Even, the largest deformations observed in nuclei are not found at the level of ground states but are manifested as excited states. Some of these excited states are termed 'superdeformed' (cf. figure 3.29). The term 'superdeformed' is used generally to imply 'more than the largest ground-state deformations observed in the rare-earth and actinide regions'. The rare-earth ground-state deformations can be viewed usefully as spheroids with approximate major:minor axis ratios of 3:2. Superdeformed structures in nuclei appear to have approximate major:minor axis ratios of 2:1; but there are no established upper limits to possible extreme deformations in nuclei.

The discovery of shape coexistence in nuclei extended over many decades with a history that can be divided into eras of 'an exotic rarity', followed by 'characterization of localized mass regions of prevalence', and culminating in 'proposing a universal feature of all nuclei'. (While nuclei with only one bound state would not be considered as candidates, among unbound resonances there are many candidate states, e.g. ^{12}C + ^{12}C resonances as manifestations of ^{24}Mg.) The history of this saga has been sketched in an essay-type publication [1], there are four review articles [2–5] and there is a focus issue of a journal [6] that documents the details as they emerged over a period of 60 years.

The protracted history for this discovery profile is significant because the key signatures of nuclear shape coexistence had to be deduced against a background of other nuclear degrees of freedom. This was not easy because quantum mechanical mixing between the coexisting shapes always occurs to varying degrees; and sometimes the 'other nuclear degrees of freedom' were ambiguous because of failure to recognize that they coexist with structures (excited states) possessing different shapes.

[2] This term is taken from chemistry where 'isomer' means 'iso = same', 'mer = units'. For example, the tastes that we call 'orange' and 'lemon' both stem from the molecule limonene which is 'chiral' (left-handed is 'orange', right-handed is 'lemon').

Figure 4.4 shows the states in even-Sn isotopes that are the lowest deformed-band members (cf. figure 2.13). Figure 4.5 shows $B(E2)$ values, where lifetime and γ-ray branching ratio data are available, that support the presence of structures with very different quadrupole collectivity. Recently, it has been shown that the structures in ^{116}Sn are fairly complex, details are given in figure 4.6. This can be understood in terms of configuration mixing. A simple view of the consequences of mixing of spherical and a deformed configuration with spin-parity 2^+ is shown in figure 4.7. The global systematics of $B(E2; 0_2^+ \rightarrow 2_1^+)$ values correlates with the $B(E2; 2_1^+ \rightarrow 0_1^+)$ values (cf. figure 2.14 and 2.15).

A subtle point to note is that $B(E2)$ values are related to $E2$ transition matrix elements by a spin factor as evident in equation (3.8); namely, there is a spin

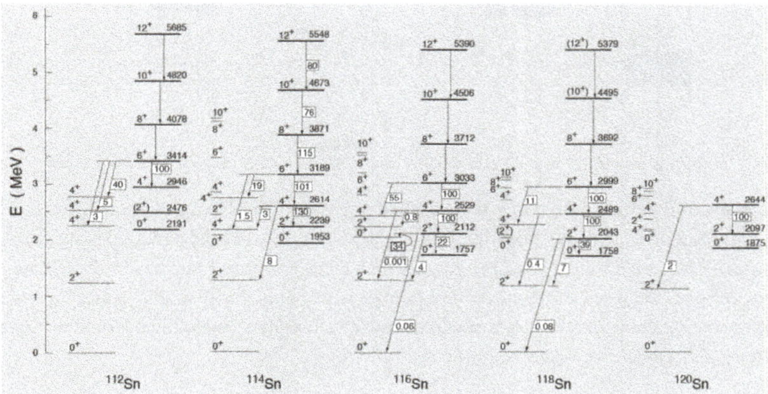

Figure 4.4. Deformed bands in the tin isotopes. The figure shows $B(E2)$ values in W.u. and, where a decay branch is indicated by '100', relative $B(E2)$ values. The figure is reproduced from [7], copyright (2010) World Scientific Publishing Company.

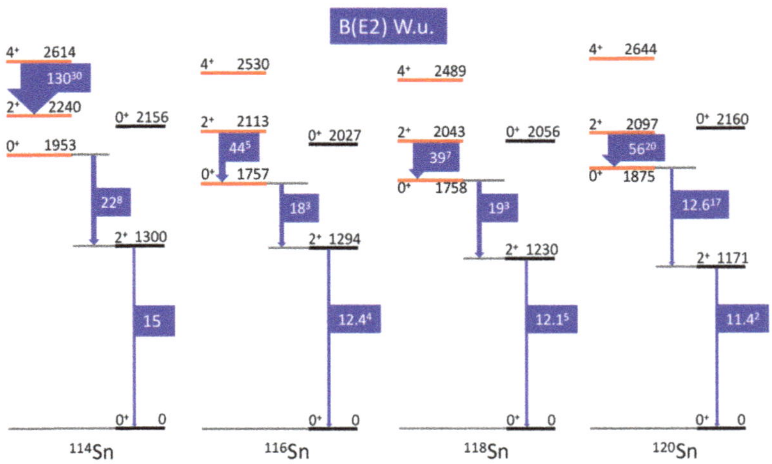

Figure 4.5. The systematics of $B(E2)$ strengths, given in W.u., for the low-lying excited states in $^{114-120}$Sn. The significant values for the B(E2; $0_2^+ \rightarrow 2_1^+$) strengths, i.e. between the deformed band heads and the spherical structures, are explained in figure 4.7.

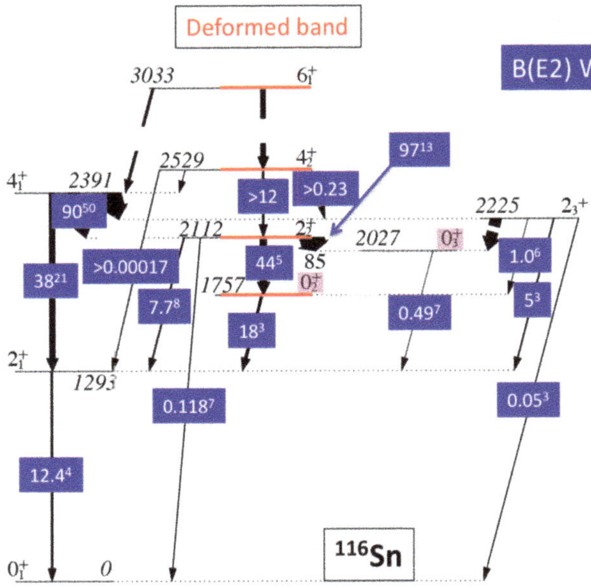

Figure 4.6. Details of the decays from low-lying excited states in ^{116}Sn. Decay of the deformed, 6_1^+ state to the spherical, 4_1^+ state and deformed, 4_2^+ state indicates mixing occurs for these 4^+ states. Mixing also occurs for the 0_2^+ and 0_3^+ states as indicated by the $B(E2)$ strength which has significant branches to these states from the 2_3^+ deformed state (also cf. the $\rho^2(E0)\cdot 10^3$ shown in figure 4.8). Probably the 2_2^+ and 2_3^+ states are also the result of mixing of the coexisting configurations: a large $E0$ transition strength between them would be expected if this is occurring. The $B(E2)$ values are taken from [8].

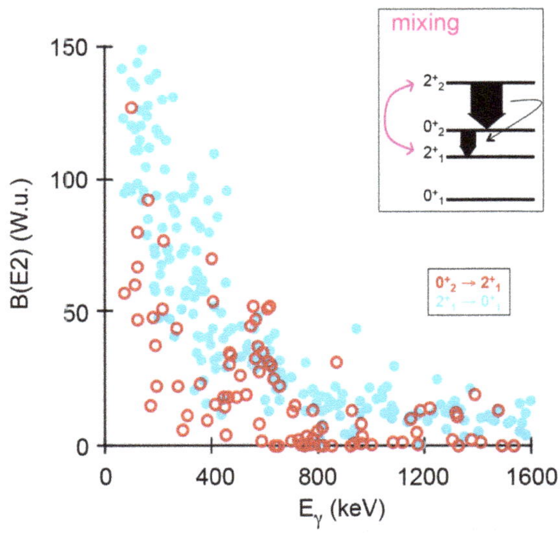

Figure 4.7. A global view of $B(E2; 0_2^+ \rightarrow 2_1^+)$ strengths versus γ-ray transition energies, shown for all nuclei. The mechanism that causes this is mixing, as illustrated in the inset, and as supported by the overlay of global $B(E2; 2_1^+ \rightarrow 0_1^+)$ strengths (cf. figure 2.14 and 2.15). Note that, due to spin factors, $B(E2; 0_1^+ \rightarrow 2_1^+)$ would be $5 \times B(E2; 2_1^+ \rightarrow 0_1^+)$.

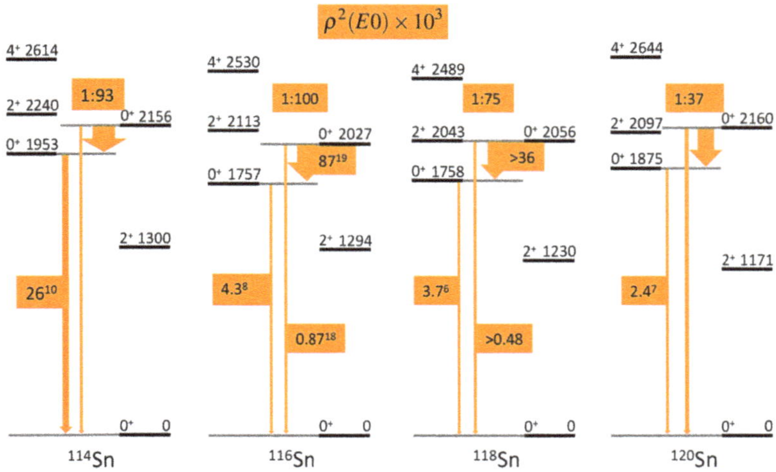

Figure 4.8. Electric monopole transition strengths, ρ^2(E0) $\cdot 10^3$ in $^{114-120}$Sn. The $0_3^+ \rightarrow 0_2^+$ transition strengths support the mixing of underlying configurations with very different mean-square charge radii, indicative of the deformation of the 0_2^+ states. See also figure 4.9 for more details of mixing.

dependence on the initial state, viz. $1/(2I_i + 1)$, which results in a given $E2$ transition matrix element leading to a $B(E2)$ value that is 5× larger for $I_i = 0$ compared to $I_i = 2$. This reflects the fact that $I = 0$ possesses only a single substate of $M = 0$, whereas $I = 2$ possesses five M substates, $M = 2, 1, 0, -1, -2$: decay $I = 0 \rightarrow I = 2$ can occur to five substates; decay $I = 2 \rightarrow I = 0$ is only to a single substate. Failure to recognize this spin factor can (and has) resulted in the mistaken view that a few Weisskopf units of $B(E2)$ strength is indicative of weak quadrupole collectivity in the $E2$ decay of excited states with $I^\pi = 0^+$.

The configuration mixing in the even-Sn isotopes also occurs between the underlying deformed 0^+ configuration and an excited 'spherical' 0^+ configuration. The evidence for this is shown in figure 4.8. This depicts the identification of electric monopole, $E0$ transition strength in $^{114-120}$ Sn. The decay of excited states in nuclei by E0 transitions has been established to be universally associated with the coexistence of configurations, with very different mean-square charge radii, which mix. Details of the mixing mechanism are presented in figure 4.9. Nuclei do not possess a low energy monopole or 'breathing' mode; but this can be effectively generated via the configuration mixing shown[3]. Although such transitions occur in closed shell nuclei, their description in shell model terms is very challenging because the model is usually confined to a single shell within which all configurations have the same mean-square charge radius. The underlying nature of the deformed configurations is revealed by the spectroscopy described below.

Figure 4.10 shows the similarities in excitation energies for the deformed band in ^{116}Sn and the ground-state bands in the isotones ^{112}Pd and ^{120}Xe, which are

[3] This was first recognized independently by John Rasmussen and by Jan Blomqvist in private communications and was formally presented by Kantele and coworkers [9]. We are indebted to Rauno Julin for this important item of the historical record.

E0 transition strengths: a model-independent description

Monopole strength parameter

$$\rho_{if}(E0) = \frac{\langle f|\sum_j e_j r_j^2|i\rangle}{eR^2} \equiv \frac{\langle f|m(E0)|i\rangle}{eR^2} \equiv \frac{M_{if}(E0)}{eR^2}$$

Monopole strength from mixing of states with different <r²>

$$|i\rangle = \alpha|1\rangle + \beta|2\rangle, \quad |f\rangle = -\beta|1\rangle + \alpha|2\rangle$$

$$M_{if}(E0) = \alpha\beta\{\langle 2|m(E0)|2\rangle - \langle 1|m(E0)|1\rangle\} + (\alpha^2 - \beta^2)\langle 1|m(E0)|2\rangle$$

$$\boxed{M_{if}(E0) \approx \alpha\beta\,\Delta\langle r^2\rangle}$$

$$\Delta\langle r^2\rangle \equiv -\langle 1|\sum_j e_j r_j^2|1\rangle + \langle 2|\sum_j e_j r_j^2|2\rangle$$

Figure 4.9. Illustration that mixing is the cause of *E*0 transition strength in nuclei. It results from the mixing of configurations with different mean-square charge radii. The quantum mechanics involved is sketched in essential terms. The strength depends on the difference in the mean-square charge radii and the degree of mixing: if there is no mixing, the strength will not be manifested.

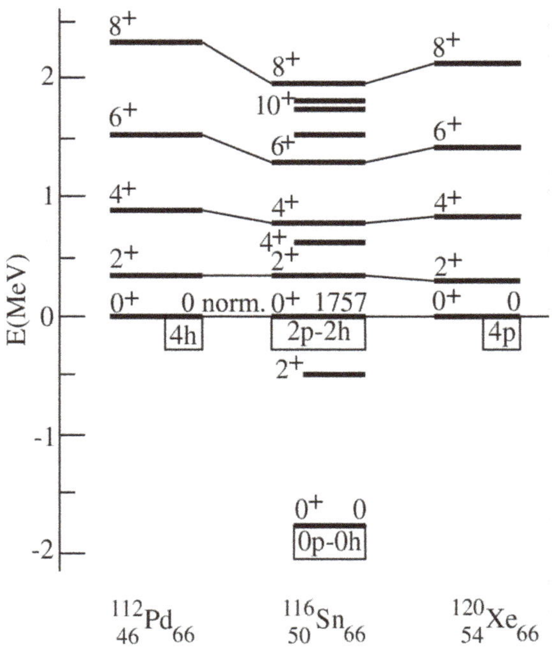

Figure 4.10. An illustration of the similarity between the deformed π(2p–2h) band in ^{116}Sn and the ground-state bands in the isotones ^{112}Pd, π(4h) and ^{120}Xe, π(4p). Reprinted with permission from [5]. Copyright (2019) by the American Physical Society.

distinguished by possessing four active protons, four 'holes' and four 'particles', respectively: the inference is that the deformed structure in ^{116}Sn is a proton two-particle-two-hole, π(2p–2h) excitation. The direct proof of this is shown in figure 4.11 (illustrated here for ^{118}Sn, which is nearly identical to ^{116}Sn, cf. figures 4.4, 4.5 and 4.8), i.e. the addition of a proton pair, effected by the reaction (^{3}He,n) on a target of ^{116}Cd, directly populates the ^{118}Sn ground state and a structure at 1.77 MeV. The energy resolution is poor because the energy analysis is for a neutron ejectile and this had to be carried out by time of flight (high-resolution magnetic analysis is unsuitable!) Thus, it appears that both the 1758 and 2056 keV 0^+ states in ^{118}Sn are populated in this process. Note, angular distributions of the ejectile neutrons establish that the final state(s) has spin zero.

The most dramatic feature of a π(2p–2h) state occurring at an excitation of 1.8 MeV in ^{116}Sn is that the π(1p–1h) states lie above 4 MeV, as observed in the one-proton transfer reaction ^{115}In(^{3}He,d)^{116}Sn, presented in figure 4.12. The implication of this is that correlation energy has lowered the π(2p–2h) structure by over 6 MeV. In shell model terms, this is an enormous energy. Further, the correlations can be deduced to produce deformation. From the 'parabolic' energy trend in these states, with minima at mid shell, it is manifestly a many-proton-pair-many-neutron-pair correlation. One does not easily obtain a direct observation of such correlation

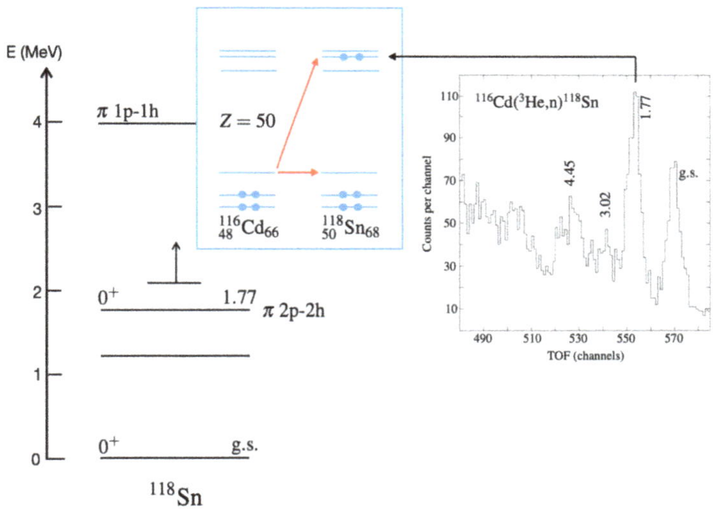

Figure 4.11. A schematic view of the two-proton stripping reaction process by which the deformed state in ^{118}Sn is populated. This supports the dominance of an underlying π(2p–2h) structure for the deformed configuration. In the inset, the time-of-flight (TOF) spectrum for neutrons in the two-proton stripping reaction ^{116}Cd(^{3}He,n)^{118}Sn. The events corresponding to the ground state are marked as g.s. All of the strong peaks corresponding to 0^+ states are shown. The peak at 1.77 MeV matches the excitation energy of the lowest energy deformed state in this isotope (cf. figures 4.4, 4.5 and 4.8). The spectrum is taken from [10] and has been redrawn and supplied, courtesy of Paul Garrett.

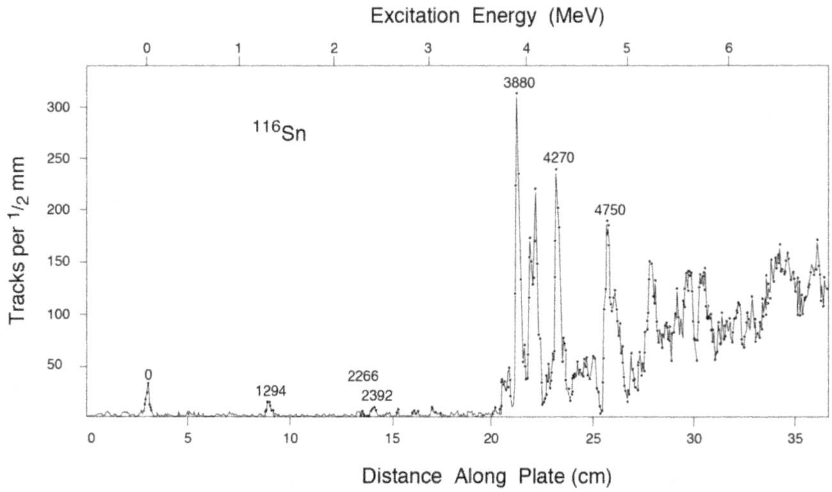

Figure 4.12. Spectrum of deuterons from the one-proton stripping reaction ^{115}In(^{3}He,d)^{116}Sn. The strength reveals the excitation at which π(1p–1h) structures become important and reflects the large shell model energy gap of ~4 MeV at $Z = 50$. In contrast, the π(2p–2h) configurations are at ~2 MeV, i.e. ~6 MeV lower than a naïve expectation that they should be at ~8 MeV: this is the result of correlation energy. The figure is reproduced from [7], copyright (2010) World Scientific Publishing Company. The original data on which this figure is based appears in [11].

energies in association with deformation because usually all of the observed states in a given nucleus have the same deformation. (We will see below that 4p–4h, even 8p–8h deformed states can occur at low energy and involve very large correlation energies, which thereby in a number of nuclei result in an ordering of levels that is not consistent with the independent single-particle shell model. This has led to a persistent misunderstanding, even confusion in the community with respect to the origin of such obervations (see, e.g. figure 4.27 for the case of ^{40}Ca).

Odd-mass nuclei adjacent to closed shells also exhibit shape coexistence. An example is shown in figure 4.13. The configurations involved are termed ‘intruder’ states, i.e. states that show up where they are not expected. They are often labeled using spherical shell model configurations: but they are not spherical shell model states; they have many characteristics of Nilsson states. Most critically, they involve strong pairing correlation contributions to their energies, which enhance the lowering of their excitation energies: this is illustrated shortly, in section 4.3 (cf.., figure 4.31). Further, they exhibit similar ‘parabolic’ energy systematics to the deformed states in the neighboring closed shell nuclei, as shown in figure 4.14. Details of these structures are beyond the present level of focus.

The most remarkable occurrence of shape coexistence in singly closed shell nuclei occurs in the $N = 20$ isotones. This is depicted in figure 4.15. The ground state of ^{32}Mg is highly deformed. The figure illustrates the probable systematic trend of the deformed states in these nuclei. It is surprising that the possibility for observation of shape coexistence in ^{36}S has, to our knowledge, never attracted attention.

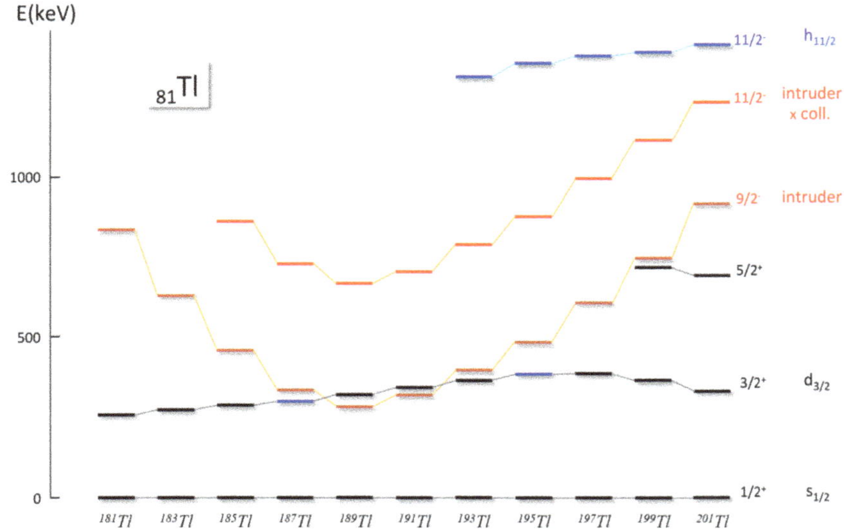

Figure 4.13. Collective bands and proton hole states in the odd-mass Tl ($Z = 81$) isotopes. The term 'intruder' is clarified in figure 4.14.

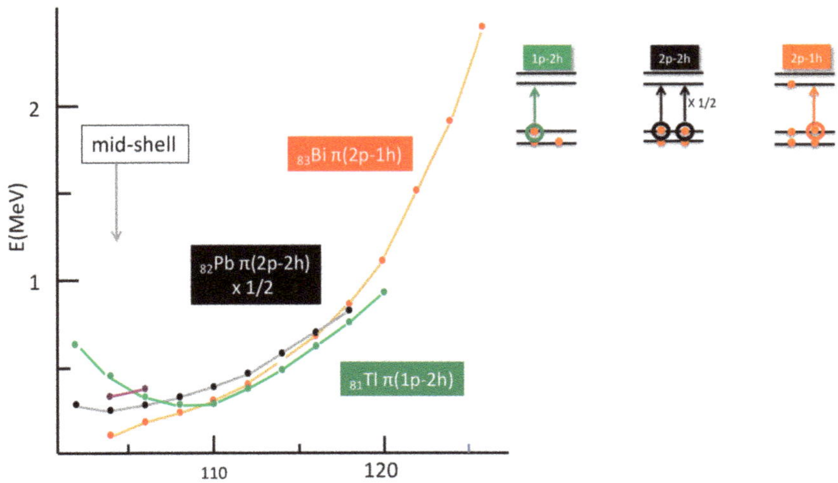

Figure 4.14. Similarity in excitation energies of deformed-band 0^+ states in the Pb ($Z = 82$) isotopes and the intruder state band heads in the odd-mass Tl ($Z = 81$) and Bi ($Z = 83$) isotopes. The structure of the intruder states is shown schematically in the top right-hand corner. They can be described as 1p–2h states in the Tl isotopes and as 2p–1h states in the Bi isotopes: the letter 'p' is the configuration of the 83rd proton, i.e. a $1h_{9/2}$ particle; the letter 'h' is the configuration of the 81st proton, i.e. a $3s_{1/2}$ hole. The correlations are manifestly produced by the '2p' and '2h' configurations, i.e. the pairs: these are correlated pairs of protons (particles and holes behave equivalently), which interact with the correlated pairs of neutrons across the shell to produce the parabolic energy trend with a minimum at mid shell, i.e. where the number of neutron pairs is a maximum. Note that the energy plotted for $_{82}$Pb π(2p–2h) is $1/2 \times E(0_2^+)$. This matches the observed systematics of the π (1p–2h) and π(2p–1h) state energies and is a clear signature that: *the energies are controlled by the number of correlated proton pairs, independent of the unpaired particle configurations.*

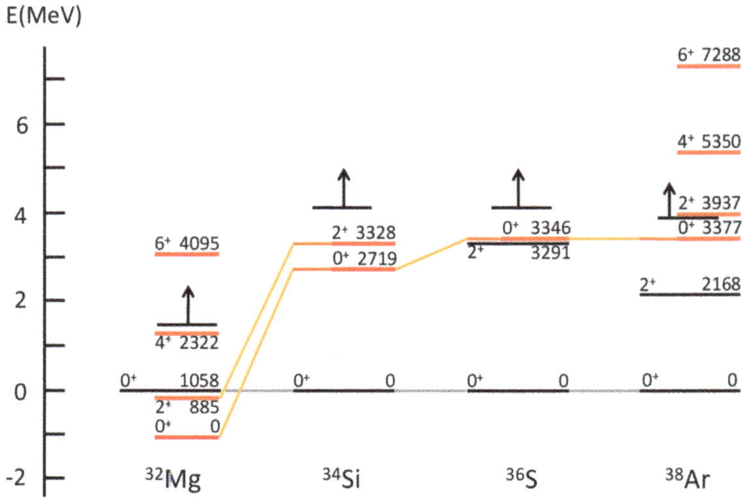

Figure 4.15. Shape coexistence and intruder states in the $N = 20$ isotones. The 0_2^+ state identifications are made in: ^{32}Mg [12] and ^{34}Si [13]. The intruder states can be understood in an exactly parallel manner to the situation in the Sn isotopes. Thus, here the 2p–2h configurations involve neutron pairs interacting with protons. The excitation pattern reflects proton subshell structure ($2s_{1/2}$, $1d_{3/2}$) as these orbitals are filled: this is beyond the present level of discussion. Note that the deformed bands in ^{32}Mg and ^{38}Ar possess nearly identical energy spacing.

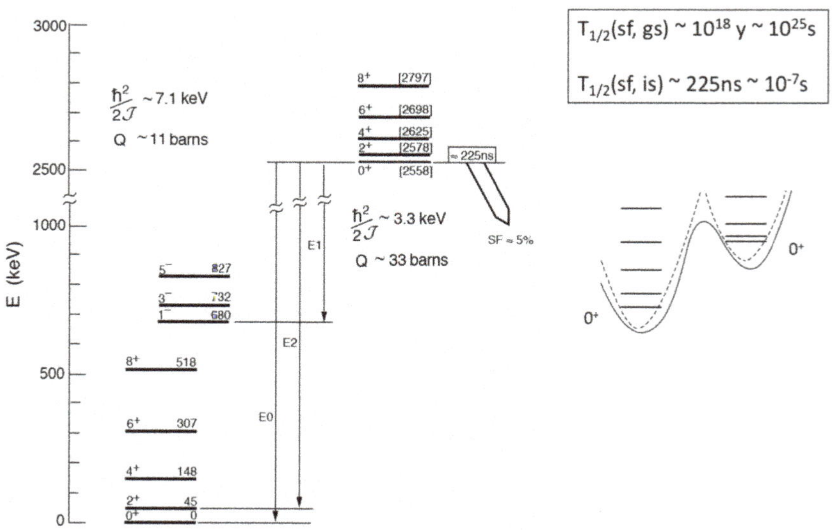

Figure 4.16. Shape coexistence in the deformed actinide region. The highly deformed structure exhibits spontaneous fission which occurs with a probability of $\sim 10^{30}\times$ faster than the spontaneous fission of the ground states of these nuclei. This is a Coulomb barrier tunnelling effect, cf. figure 1.18. The figure is reproduced from [7], copyright 2010 World Scientific Publishing Company.

Shape coexistence also occurs in a number of circumstances that indicate the need for a much wider framework of search and description. Figure 4.16 illustrates an example of the occurrence of a superdeformed band in a nucleus with a deformed ground state. Figure 4.17 illustrates, perhaps the most dramatic of all occurrences of

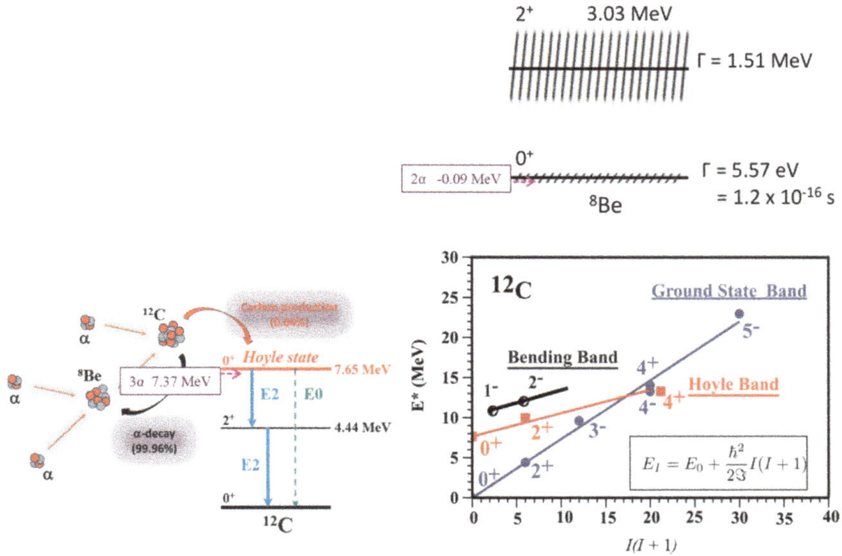

Figure 4.17. Details of the so-called 'Hoyle state' in ^{12}C—its formation and decay—this is a highly deformed state in a nucleus with a moderately deformed (oblate) ground state. If this state did not exist, 'we would not be here'—see [14]. The use of cross-hatching of levels is to indicate, semi-quantitatively, those levels with energy widths, which are given by Γ. An energy width, Γ is related to a lifetime, τ by equation (2.1) with ΔE replaced by Γ, i.e. $\Gamma = \tau\hbar$ (this equivalence, in practical units, is shown for the ground state of ^{8}Be). Such states are called resonances: they are unbound and they are not energy eigenstates (although here they are depicted as such). Recall, an energy eigenstate is a so-called 'stationary' state and is characteristic of a bound state of a system. If the Hamiltonian of a bound system is augmented by the Hamiltonian of the electromagnetic field, then excited states are no longer 'stationary': they decay by an electromagnetic process, e.g. photo emission (and the excited state can be described by a level width). (See [15, 16] for new measurements of the $E2$ and $E0$ radiative widths of the Hoyle state.) Here, the states (resonances) in ^{8}Be are unbound with respect to decaying into two ^{4}He nuclei. Quantum mechanically, the 'four-proton-four-neutron' system can exist in two low-energy states $|^4\text{He} + ^4\text{He}\rangle$ and $|^8\text{Be}\rangle$. One could describe this system as resonating back and forth between the two states in a high-temperature (> 0.090 MeV) environment such as occurs in massive stars. The quantum mechanics needed to understand this is introduced in [17], chapter 9, especially sections 9.10 and 9.11. Reprinted with permission from [18]. Copyright (2014) by the American Physical Society.

shape coexistence, the Hoyle state in ^{12}C—this is a highly deformed state coexisting with a ground state of moderate deformation—and is the 'gateway' to stellar synthesis of elements heavier than helium.

4.2 Pairing-dominated structures in single-closed shell nuclei

The most easily identified structures in singly closed shell even-mass nuclei are illustrated in figure 4.18 and 4.19. The excitation patterns can be understood from the semi-classical perspective of orbital 'overlap' and a short-ranged attractive pairing interaction, as illustrated in figure 4.20. The structures presented in figure 4.18 and 4.19 are dominated by $1h_{11/2}$ shell model configurations, neutron configurations in the Sn isotopes and proton configurations in the $N = 82$ isotones.

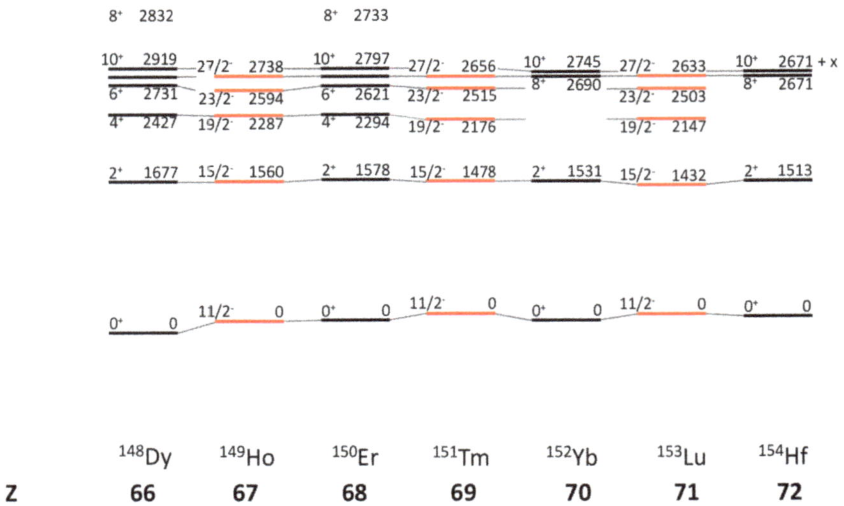

Figure 4.18. Singly closed shell nuclei with $N = 82$ for which the low-energy structure is dominated by a broken pair in the $1h_{11/2}$ orbital. Note, in ^{152}Yb and ^{154}Hf the decay of the 8^+ states (not shown in the figure) is to 7^- states at 2550 and 2457 keV, respectively; in ^{148}Dy and ^{150}Er the 8^+ states decay partially to 7^- states at 2738 and 2633keV, respectively. Other states at low energy are 3^- states—^{148}Dy(1688), ^{150}Er(1786), ^{152}Yb (1890), ^{154}Hf(2011), and 5^- states—^{148}Dy(2348), ^{150}Er(2260), ^{152}Yb(2203), ^{154}Hf(2146). These negative-parity states involve a single broken pair, either $1h_{11/2}$-$3s_{1/2}$ or $1h_{11/2}$-$2d_{3/2}$.

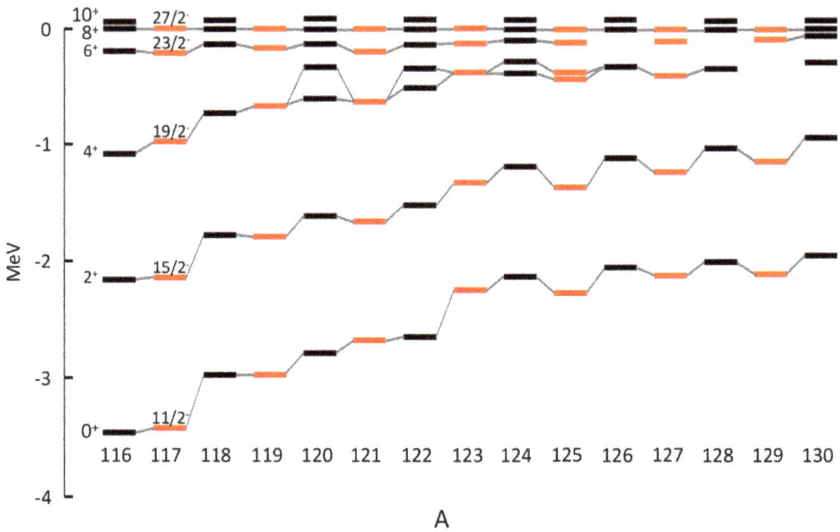

Figure 4.19. Singly closed shell nuclei with $Z = 50$ (Sn isotopes) for which the low-energy structure is dominated by a broken pair in the $1h_{11/2}$ orbital. Note, in the heaviest Sn isotopes the 8^+ states decay to 7^- states: at 1947 keV (^{130}Sn, a 100% β-decaying isomer) and see [19] for details of $^{118-128}$Sn. Details of the $23/2^-$ state decays in ^{125,127,129}Sn are given in [20]. Note some multiple copies of $J = 4$, 19/2.

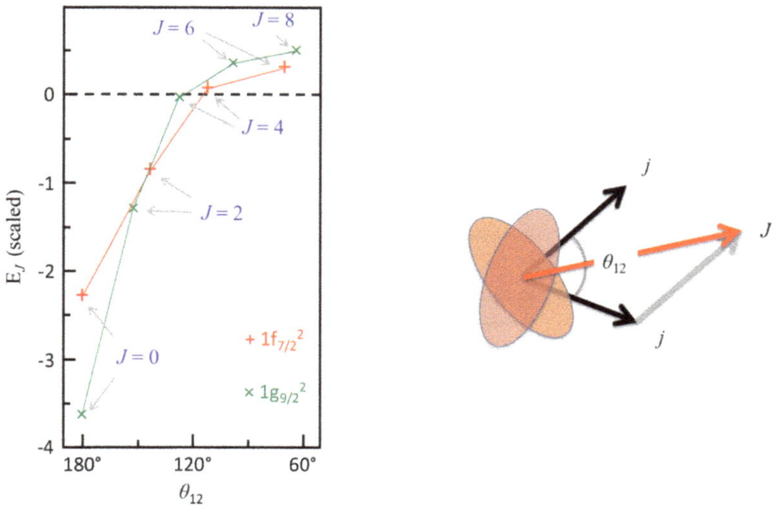

Figure 4.20. Schematic view of the interaction energy between a pair of identical high-j nucleons. The data are taken from [21] where more details can be found.

Note that the resulting states are near identical, at a semi-quantitative level, i.e. excitation energies, independent of whether protons or neutrons are involved and independent of the number of nucleons in the $1h_{11/2}$ orbital.

The states with the highest spins, shown in figure 4.18 and 4.19 are dictated by the Pauli exclusion principle: thus, for a $j = 11/2$ broken pair in an even–even nucleus, $m_j = 11/2$ and $9/2$ couple to give $M = 10$, whence $J_{max} = 10$; similarly in an odd-mass nucleus, $m_j = 11/2$, $9/2$, and $7/2$ couple to give $M = 27/2$, whence $J_{max} = 27/2$. The excitations in each nucleus are arbitrarily normalized at $J = 8$ (even mass) and $J = 27/2$ (odd mass) to reveal the similarities of the high-spin state relative energies. The even–even nuclei only exhibit even-integer spin values, again, because of the Pauli exclusion principle. This is determined using the so-called 'm scheme', details of which are given below. The possible odd-mass spin values are more complicated and can also be determined using the m scheme.

The structures behind the excitation patterns in figure 4.18 and 4.19 are dominated by nucleons occupying, predominantly a single $j = 11/2$ subshell. The pattern undergoes a simple modification if occupancy extends to a second subshell, with $j = 1/2$, as shown in figure 4.21. Basic details are presented in chapter 5, section 5.2. Manifestly, this is dominated by the mixing of two near-degenerate 0^+ configurations. Note that in even–even nuclei, with the 'depression' of the ground state, the configuration mixing leads to the 2_1^+ state appearing anomalously high. It can be tempting (some people have been tempted) to interpret a relatively high energy 2_1^+ state as the signature of an energy gap between subshells, when in fact the subshells are nearly degenerate in energy as illustrated here. In figure 2.8 some exceptionally high energies for 2_1^+ states in ^{90}Zr and ^{96}Zr are noted, also such a feature is associated with ^{68}Ni: these are all examples of the effect depicted in figure 4.21.

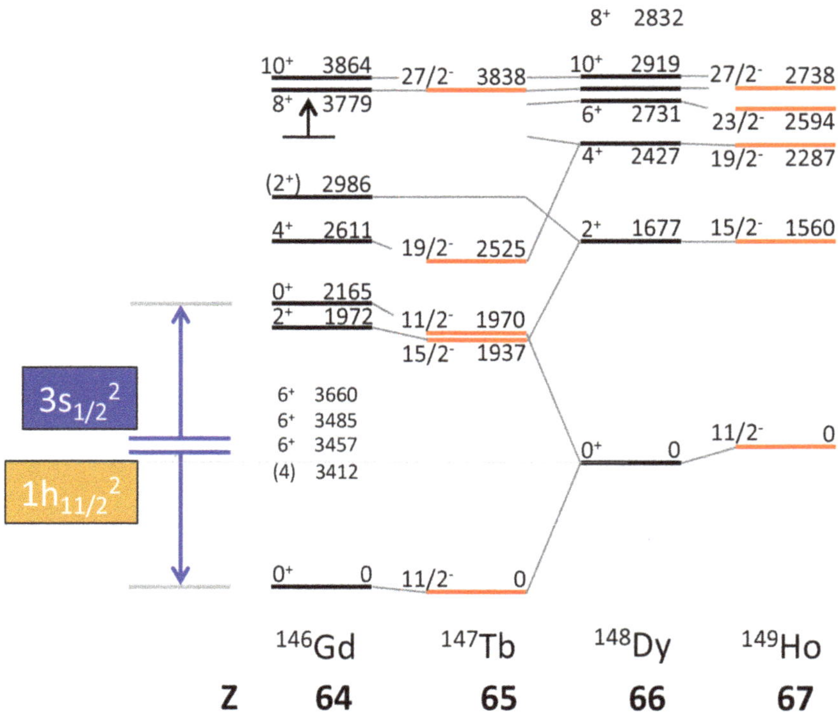

Figure 4.21. Examples of excitation spectra involving a high-j configuration, in the present case $1h_{11/2}$, i.e. $j = 11/2$, that are close in energy to a $j = 1/2$ configuration (^{145}Gd and ^{147}Tb). Such structures are discussed in detail in section 5.2.

Great caution is needed when making model-based interpretations of data: the result can be a spurious interpretation of the structure observed, when the effect is due to completely different physics than that expressed in the model. In the present circumstance, the physics of the shell model is that of independent particles confined in a spherical mean-field potential, whereas the actual physics is that of correlated pairs. These correlations are essential to making the step from a highly idealized set of independent-particle shell model states, which are a model concept, into understanding the states actually observed in nuclei. The message is: assume correlations are always present in quantum many-body systems; even when independent-particle behavior appears to dominate.

An indirect and subtle effect of pairing correlations, yet familiar in data plots, is the smooth variation of separation energies across multiple subshells that are not degenerate in energy. The naïve expectation versus the observed behavior is depicted schematically in figure 4.22. The observed behavior of S_n in the tin isotopes and of S_p in the $N = 82$ isotones, are shown in figure 4.23 and 4.24.

The lowest lying states in odd-mass single-closed shell nuclei can be assigned spherical shell model labels, again with the provison that pairing correlations play an important role in their energies. The example of the odd-Z isotones with $N = 82$ are shown in figure 4.25(a) and (b). Indeed, proton-pairing correlations will be present in all

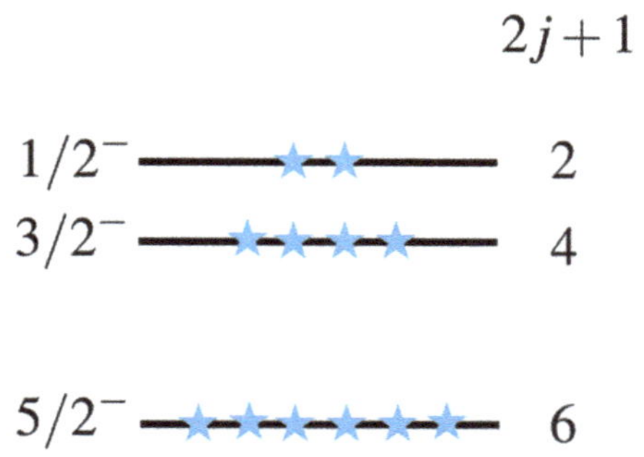

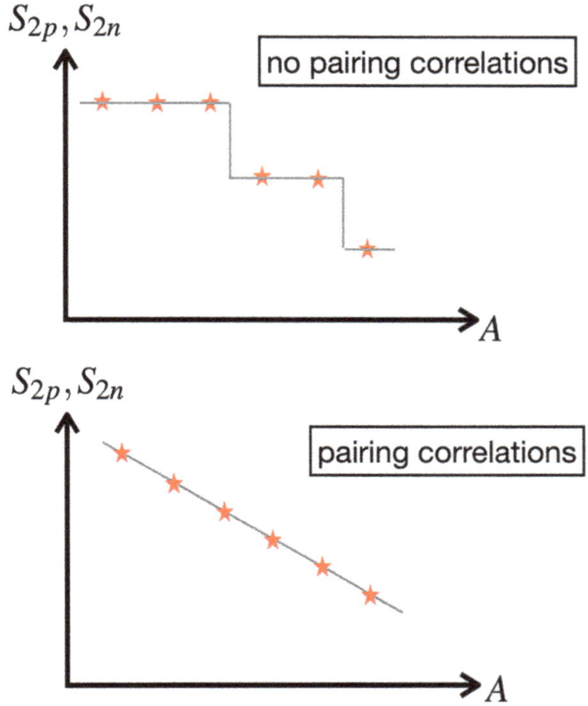

Figure 4.22. Effect of pairing correlations on two-nucleon separation energies.

of these nuclei except ^{133}Sb. There would appear to be a subshell energy gap between the $1g_{7/2}$ + $2d_{5/2}$ pair of subshells and the $1h_{11/2}$ + $2d_{3/2}$ + $3s_{1/2}$ group of subshells, as manifested in the shifts of states between ^{145}Eu and ^{147}Tb. However, a cautionary observation has already been made with respect to figure 4.21. These apparently spherical independent-particle degrees of freedom involve significant pairing correlation contributions to their energies. Discussion of this is beyond the present level of detail. There are noticeable 'extra' states in ^{135}I and ^{137}Cs, evident in figure 4.25 (b), between 800 and 1500 keV. These are the broken-pair states that can occur for the $1g_{7/2}$ configuration. Again, discussion is beyond the present level of detail.

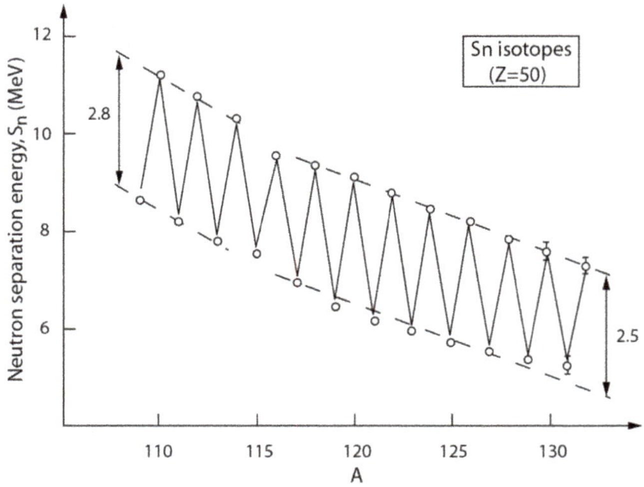

Figure 4.23. One-neutron separation energies for the Sn ($Z = 50$) isotopes. Note that the systematic trend is smooth, i.e. there is no evidence of subshells filling, cf. figure 4.22. The figure is reproduced from [7], copyright 2010 World Scientific Publishing Company.

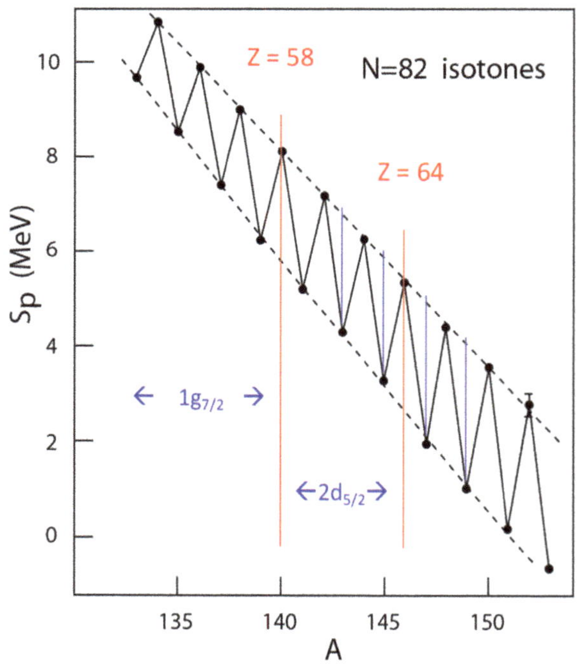

Figure 4.24. One-proton separation energies for the $N = 82$ isotones. The expected filling of the $1g_{7/2}$ and $2d_{5/2}$ subshells is shown, cf. figure 4.25(b). The figure is reproduced from [7], copyright 2010 World Scientific Publishing Company.

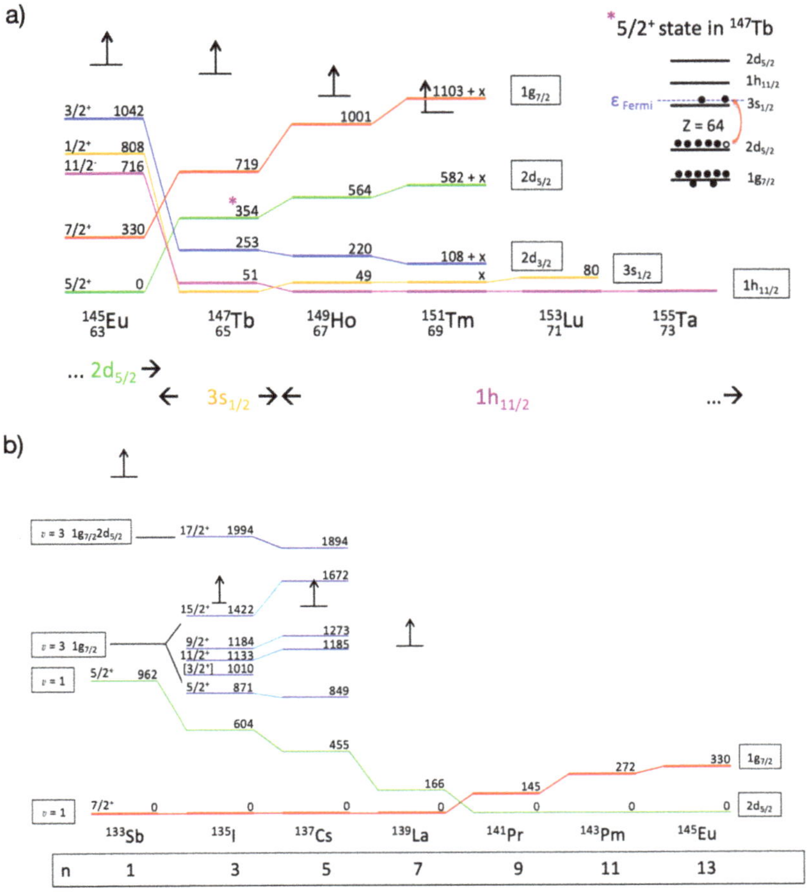

Figure 4.25. (a) and (b) Systematics of excited states in the odd-mass $N = 82$ isotones. See text for a discussion of the details. Candidate states for all of the expected shell model configurations for $50 < Z < 82$ are observed. Additional states are shown in ^{135}I and ^{137}Cs: these result from 'breaking' a pair in the $1g_{7/2}$ orbital. The so-called 'seniority' quantum number, v is used to label states, where seniority equals the number of unpaired nucleons (protons).

The nuclear structure involved in all of these excitation patterns involves so-called 'Cooper' pairs. These are highly correlated nucleon-pair degrees of freedom, which can be expressed, most simply, using the form

$$|\text{one pair in many-j subshells}\rangle = \sum_j \alpha_j |j, \bar{j}\rangle, \tag{4.1}$$

where,

$$\begin{aligned} |j, \bar{j}\rangle &:= \sum_{m=1/2}^{j} |j, +m, j, -m\rangle \\ &= \frac{1}{\sqrt{j + 1/2}} \{|+j, -j\rangle + |+j - 1, -j + 1\rangle + \cdots + |+1/2, -1/2\rangle\}, \end{aligned} \tag{4.2}$$

and the α_j are amplitudes that are related to the occupancies of the subshells, labeled by their j values.

One observes that a paired structure is 'averaged' or 'smeared' over many configurations. The amplitude α_j for the pair occupying the subshell labeled by j is observed to be a slowly varying function of the energies of the subshells. Multiple pairs are described by such expressions with α_j values that reflect the progressive filling of the energy-ordered subshells. This is the reason that sudden changes, e.g. in S_n or S_p are not observed, except at closed shells. It is the reason that S_{2n} and S_{2p} values exhibit smooth systematics trends between major closed shells (cf. figure 1.13). The sudden changes at closed shells show that the pairing force is not strong enough to smear out the large energy gaps at the major shell closures. (If this were not true, it would have been much harder to arrive at the concept of a role for a shell model in nuclei.) Some further details of Cooper pairs are given in chapter 5.

A powerful 'calculus' for determining allowed spin values for finite many-particle systems is the so-called 'm scheme'. The m-scheme method can be applied to many-fermion or many-boson systems. The key distinction is that the total wave function for a many-boson system must be symmetric and that for a many-fermion system must be antisymmetric. The key ingredient is the expression of the many-body system using eigenstates of spin-angular momentum, e.g. $|jm_j\rangle$, $j = 1/2, 3/2, 5/2, \ldots$, $m_j = +j, +j-1, +j-2, \ldots, -j+1, -j$ for fermions. We give a few examples that contain all of the key features for working with the m-scheme calculus.

Exercise 1. $(j = 3/2)^2$, *i.e. a pair of identical fermions in a spin-3/2 shell-model subshell such as* $1d_{3/2}$ *or* $2p_{3/2}$

States will be of the form $|jm_1; jm_2\rangle$, where $j = 3/2$, $m_1, m_2 = +3/2, +1/2, -1/2, -3/2$, and $M := m_1 + m_2$, viz.

m_1	m_2	M
+3/2	+1/2	+2
+3/2	−1/2	+1
+3/2	−3/2	0
+1/2	−1/2	0
+1/2	−3/2	−1
−1/2	−3/2	−2

The following rules have been used:

- +3/2 +3/2—disallowed because two identical fermions cannot occupy identical quantum states (Pauli exclusion principle);
- +1/2 +3/2—disallowed because it is indistinguishable from +3/2 +1/2, which dictates that the state must take the form

$$\frac{1}{\sqrt{2}}\{|m_1 = +3/2, m_2 = +1/2\rangle - |m_1 = +1/2, m_2 = +3/2\rangle\},$$

  i.e. an antisymmetric form under particle interchange.

These rules are applied to all possible pairwise combinations in the list.

Then, inspection of the list of M values, starting with the maximum value, indicates that spin 2 must exist: spin 2 possesses $M = +2, +1, 0, -1, -2$. Then, of the six M values, only an $M = 0$ remains, yielding a spin 0. Thus, $J = 0, 2$ exhausts the possibilities and one concludes that these are the allowed ways in which two identical spin-3/2 (fermions) can couple. Note, we do not specify which of the $|+3/2, -3/2\rangle$ and $|+1/2, -1/2\rangle$ combinations belongs to $J = 0$ and $J = 2$: the answer is that one takes linear combinations and $J = 2$ is the symmetric combination, $J = 0$ is the antisymmetric combination.

Exercise 2. $(j = 3/2)^3$
Proceeding directly using the above rules, viz.

m_1	m_2	m_3	M
+3/2	+1/2	−1/2	+3/2
+3/2	+1/2	−3/2	+1/2
+3/2	−1/2	−3/2	−1/2
+1/2	−1/2	−3/2	−3/2

$J = 3/2$, only: this is a remarkable result—$(j = 3/2)^3$ behaves as if it were $j = 3/2$; and this is often interpreted as $(j = 3/2)^3 \equiv (j = 3/2)^{-1}$, i.e. three particles in a $j = 3/2$ orbit is equivalent to one hole in a $j = 3/2$ orbit.

Exercise 3. $(l = 1)^2$, *i.e. a pair of identical bosons in a spin-1 orbit*
Proceeding, with the recognition that two identical bosons can occupy states with identical quantum numbers, but must be in symmetric combinations, viz.

m_1	m_2	M
+1	+1	+2
+1	0	+1
+1	−1	0
0	0	0
0	−1	−1
−1	−1	−2

whence, $L = 0, 2$ (note that $|m_1 = 0, m_2 + 1\rangle$ is disallowed because it must appear in a symmetric combination with $|m_1 = +1, m_2 = 0\rangle$).

Recall, mathematically, quantum states cannot be described such that more information is implied than exists. Thus, the labels used for the particles above, which are for bookkeeping purposes, must be put into a form such that exchange of the identical particles yields the same quantum state, except for a sign change (antisymmetric) for the fermions or no sign change (symmetric) for the bosons.

4.3 Doubly closed shell nuclei

Doubly closed shell nuclei are very rare. They can be usefully divided into two groups: $N = Z$ and $N > Z$. The $N = Z$ doubly closed-shell nuclei, ^{16}O, ^{40}Ca, and ^{56}Ni possess low-lying deformed states that coexist with spherical ground states. The $N = Z$ nucleus, ^{4}He, does not possess bound excited states; and excited states in ^{100}Sn are not yet explored. The $N > Z$ doubly closed-shell nuclei for which there are extensive data are ^{48}Ca, ^{132}Sn, and ^{208}Pb. We note the expectation that ^{28}O would be observed as a candidate doubly closed-shell nucleus is not realized: it is unbound in its ground state; but bound states in ^{78}Ni are just beginning to be explored [22]. The $N < Z$ nucleus, ^{48}Ni has no known excited states.

The lowest excited states in ^{48}Ca, ^{132}Sn, and ^{208}Pb are shown in figure 4.26. The first feature to note is the large energy gap due to the closed proton and closed neutron shells. Excitations appear across these gaps as expected 1p–1h configurations and, evidently, weakly collective 3^- and 2^+ states. The collectivity is manifested in $B(E3)$ and $B(E2)$ values.

The lowest excited states in ^{16}O and ^{40}Ca are shown in figure 4.27. The observation of shape coexistence in the $N = Z$ doubly closed shell nuclei, as manifested, e.g. in the 8p–8h state in ^{40}Ca, is explained by correlation energies of multiple proton pairs and neutron pairs that can be as large as 50 MeV (where the 1p–1h excitations across the shell gaps at $Z = 20$ and $N = 20$ are ~6 MeV).

The idea of shape coexistence had its origins in light nuclei. The origination of the idea was embodied in an explanation of the first excited state of ^{16}O by Morinaga [23] in 1956.

The reason that the saga of shape coexistence was written mainly in the study of heavy nuclei is likely that heavy nuclei provide long chains of isotopes that can be

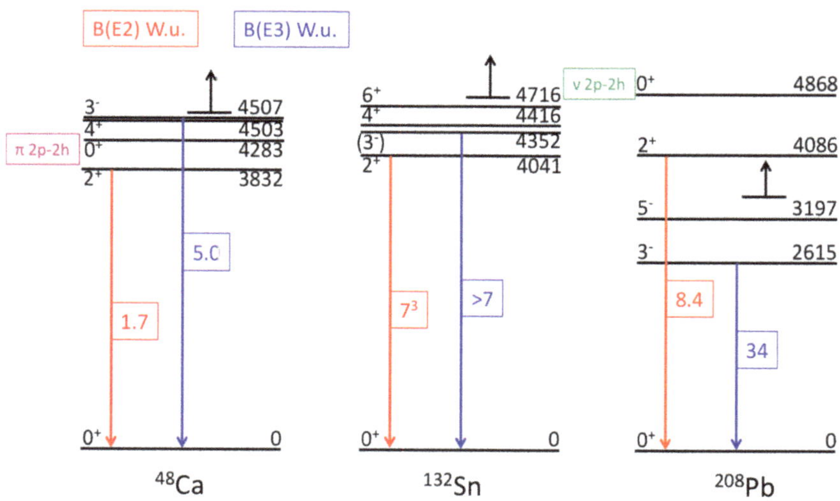

Figure 4.26. Excited states in the double-closed shell nuclei with $N > Z$. Some details of the collectivity of the 2_1^+ and 3_1^- states are given, namely electromagnetic decay strengths. Also, pair excitations are indicated where known (cf. figure 6.18).

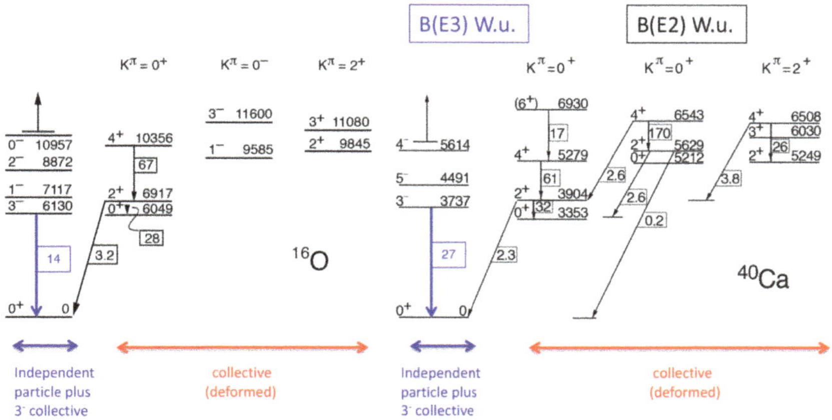

Figure 4.27. Excited states in the double-closed shell nuclei with $N = Z$. Some details of the collectivity of the 2_1^+ and 3_1^- states are given, cf. figure 4.26. The $B(E2)$ values that support collectivity are shown. Band structure that is deduced from more detailed spectroscopy, e.g. as presented in figure 6.20, is indicated with inferred K quantum numbers. Thus, the 0^+ state at 6049 keV in ^{16}O is interpreted as a 4p–4h state. A similar interpretation is made for the 3353 keV state in ^{40}Ca. The 0^+ state in ^{40}Ca at 5212 keV is interpreted as an 8p–8h state: details are beyond the present scope for discussion. Reproduced from [7], copyright (2010) World Scientific Publishing Company.

studied systematically and in detail, whereas light nuclei only have short isotopic and isotonic chains accessible to detailed study. In short chains, changes are so sudden that patterns can be difficult to discern.

Nuclei adjacent to doubly closed shells exhibit excitations that are closest to the ideal spherical shell model independent-particle (or hole) states. Examples are shown in figure 4.28. However, great caution is needed when applying this view to all nuclei adjacent to doubly closed shells. An example of a dramatic departure from the shell model view is manifested in ^{17}O where the low-energy states, while naively interpreted as one-neutron independent particle states, exhibit a quadrupole collectivity characteristic of a particle carrying an electric charge of $+0.5e$. The data are shown in figure 4.29. The structures cannot be simple one-neutron configurations, although one can find the states in ^{17}O being used as illustrations of independent-neutron degrees of freedom in the $N = 9$ isotones in various published works. Even in the ground-state structure of ^{17}O, there must be significant contributions of proton particle–hole configurations to explain the $E2$ properties of this nucleus.

Nuclei adjacent to ^{208}Pb exhibit excitations that are a rare example of so-called ‘weak coupling’. This is illustrated in figure 4.28. The ‘core’ excitations in ^{208}Pb, notably the 3^- state at 2614 keV, gives rise to closely spaced multiplets in all of the odd-mass neighbors (figure 4.30). The centroids of the energies of these multiplets are very close to the energy of the 3^- state in the core, hence the description as ‘weak coupling’. The spins of the multiplets are simply the naïve vector coupling of the spin of the odd particle or hole with the $J = 3$ core excitation. Parities are determined by multiplication, e.g. minus times minus equals plus.

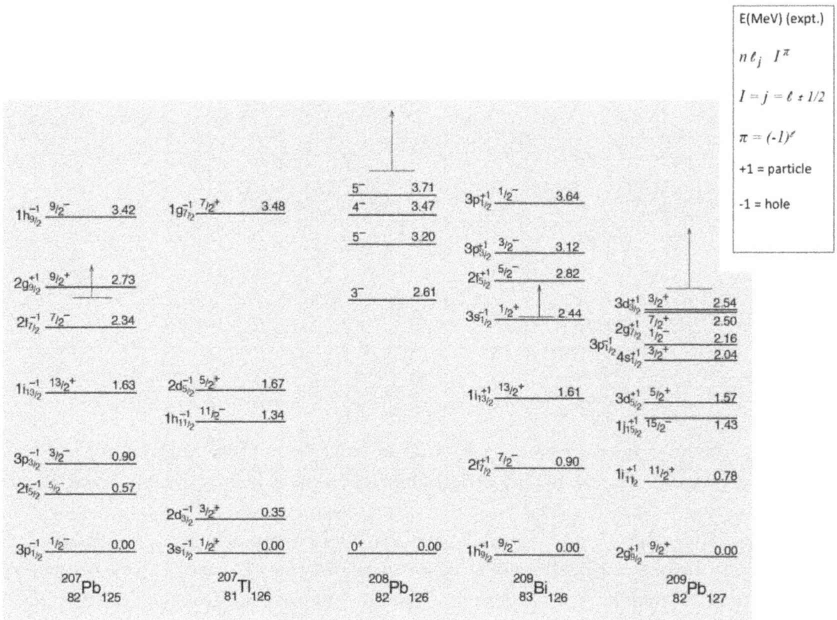

Figure 4.28. Extreme independent-particle and -hole view of odd-mass nuclei adjacent to ^{208}Pb. Candidate states for all of the expected shell model particle (^{209}Bi, ^{209}Pb) and hole (^{207}Tl, ^{207}Pb) configurations are observed. Some particle states in ^{207}Pb and some hole states in ^{209}Bi and ^{209}Pb are also shown. The low energies of these states needs explanation, which is given in figure 4.31. (A set of video-based exercises on deducing the nuclear structure of ^{207}Pb is presented in exercise 4-11.) Spectroscopic evidence for the particle and hole states in ^{209}Bi are presented in figures 6.14 and 6.15, respectively. Reproduced from [7], copyright (2010) World Scientific Publishing Company.

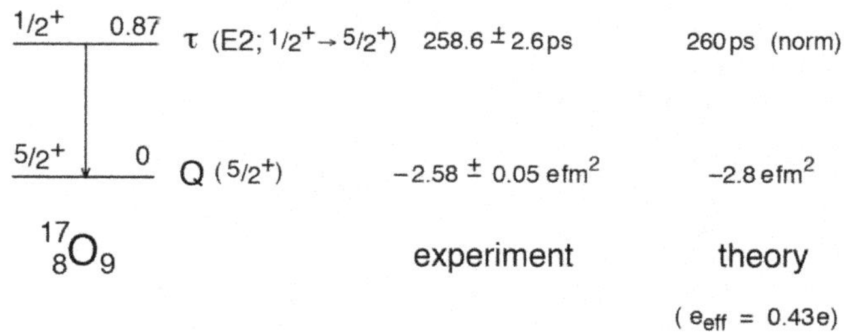

Figure 4.29. Evidence for the strong collectivity in ^{17}O involving the ground state. Some more details are given in the text. Reproduced from [7], copyright (2010) World Scientific Publishing Company.

The odd-neutron nuclei adjacent to ^{48}Ca, namely ^{47}Ca and ^{49}Ca, exhibit another example of correlation energy causing very large shifts in the excitation energy of certain configurations. Figure 4.31 shows what appears to be a collapse of the $N = 28$ shell gap, in the observation of a $3/2^-$ excited state in ^{47}Ca at 2.01 MeV and the observation of a $7/2^-$ excited state in ^{49}Ca at 3.36 MeV. The figure presents the data that exclude collective degrees of freedom from consideration in the interpretation of these states and shows that they have large pairing correlation energy contributions, i.e. they are neutron 1p–2h (^{47}Ca) and 2p–1h (^{49}Ca) states.

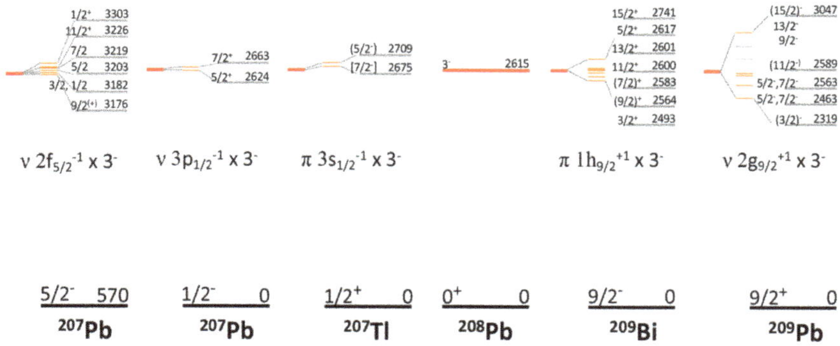

Figure 4.30. Weak coupling in odd-mass nuclei adjacent to ^{208}Pb. The 'parent' state for these structures is the 3^- state in ^{208}Pb at 2615 keV. This is a collective 'octupole' excitation as indicated by a strong $B(E3)$ decay, cf. figure 4.26; and as supported by inelastic scattering data, cf. figure 6.17. Inelastic scattering data for ^{209}Bi, cf. figure 6.16, confirm this view of collective character. (A set of video-based exercises on deducing the nuclear structure of ^{207}Pb is presented in exercise 4-11.)

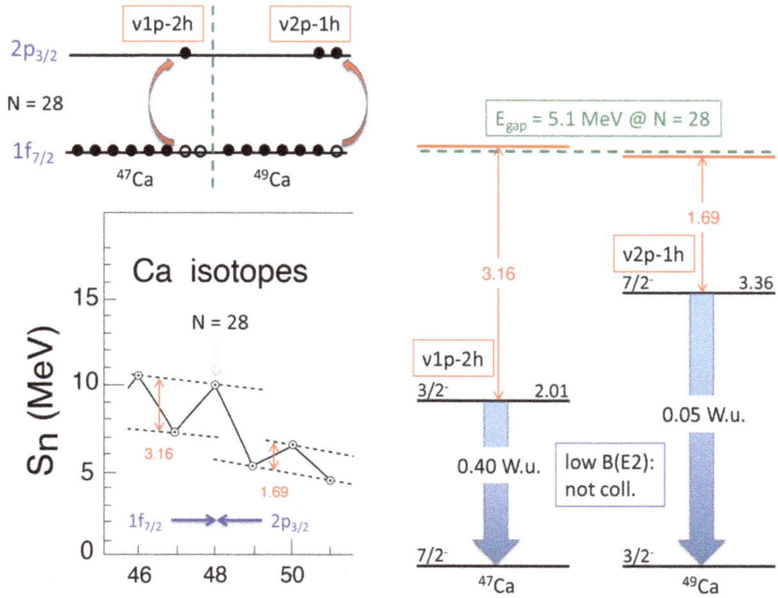

Figure 4.31. Intruder states in ^{47,49}Ca. The data for ^{49}Ca are taken from [24]. The unexpectedly low energy of the $3/2^-$ state in ^{47}Ca and the $7/2^-$ state in ^{49}Ca, which are not collective excitations of the ground states in these nuclei (as indicated by the very small $B(E2)$ values), can be understood as resulting from pairing correlations. The pairing correlation energy gain is explained using the information on the left-hand side of the figure. Thus, from a direct adoption of the odd–even staggering of S_n values, pairing correlation energies of 3.16 and 1.69MeV, respectively, lower the energies of the 1p–2h state in ^{47}Ca and the 2p–1h state in ^{49}Ca by these amounts from an 'uncorrelated' energy of 5.1 MeV.

4.4 Weakly deformed nuclei

We use the terminology ‘weakly deformed nuclei’ as a description of the nuclei that so far have not received attention with respect to recognizable patterns of excited states.

Figure 4.32 shows the yrast states in the even-mass Te ($Z = 52$) isotopes, which are adjacent to the closed-shell Sn isotopes. A number of features should be noted. First, the nucleus ^{134}Te possesses a closed neutron shell ($N = 82$) and is dominated by a pairing interaction acting on the pair of protons in a $1g_{7/2}$ configuration (cf. figure 4.20). The characteristic close energy spacing, indicative of a pairing-interaction-dominated structure, persists for some higher-spin states, notably for states with spin 4 and spin 6 in ^{130,132,134}Te and for states with spin 8 and spin 10 in $^{126,128,\,130,132}$Te. The latter, higher-spin states involve pairs that are dominated by a pairing interaction acting on a pair of neutrons in a $1h_{\ 11/2}$ configuration (cf. figure 4.19). Possibly, a pairing-dominated structure is evident in ^{106}Te and may be due to a proton-pair or a neutron-pair $1g_{7/2}$ structure.

Second, $B(E2)$ values for transitions between the yrast states exhibit a steady increase in quadrupole collectivity towards mid shell ($N = 66$). These details can be compared with the global map of $B(E2;\, 2_1^+ \rightarrow 0_1^+)$ values for this mass region, shown in figure 2.14. However, the data are incomplete; even, large fluctuations suggest that some of these numbers are unreliable (considering the basic pattern in figure 2.14).

An interpretation of the energy patterns in terms of a simple rotor or vibrator model can be tempting. Figure 4.33 shows the systematic occurrence of a low-energy 2_2^+ state in the Te isotopes, with possible band members (viz., spins 3, 4 and a few higher spins in ^{114,116,118}Te). Proof of a collective vibrational structure necessitates detailed information on many non-yrast states, which are lacking for these isotopes. Chapter 5 presents the main features that are exhibited by a harmonic quadrupole

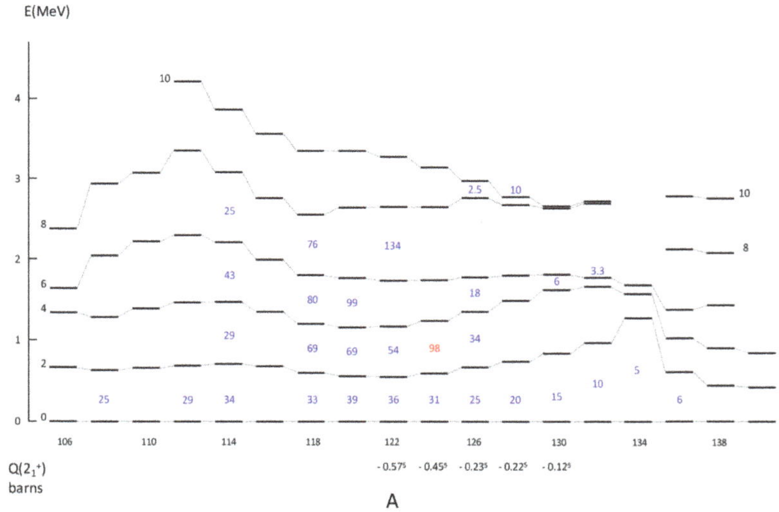

Figure 4.32. Yrast systematics in the even-mass Te ($Z = 52$) isotopes. Transition $B(E2)$ values, where measured, are shown in blue between levels. Quadrupole moments of 2_1^+ states, where measured, are shown below the isotope mass numbers. The $B(E2)$ value highlighted in red would appear to be anomalous and a re-measurement is in order. Some more details are given in the text.

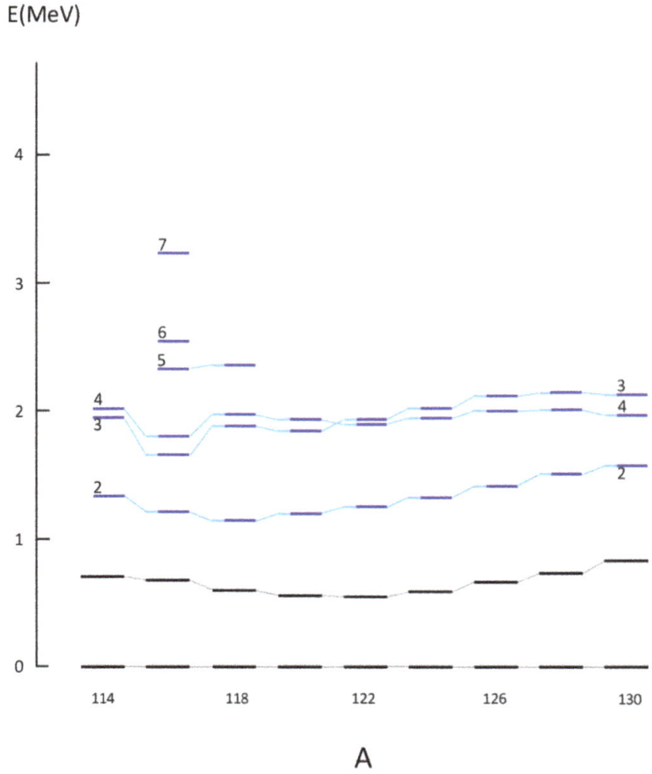

Figure 4.33. Systematics of bands built on top of 2_2^+ states in the even-Te ($Z = 52$) isotopes. Some more details are given in the text.

vibrator. A survey of such criteria, with a critical analysis, is beyond the level of the present treatment; however, we note that there are no convincing examples of such behavior in any nucleus. We further note that in the Te isotopes, where there are $Q(2_1^+)$ data (shown in figure 4.32), the indication is that significant deformation occurs towards mid shell.

The process of inspection for even-mass yrast states can be continued into the Xe ($Z = 54$), Ba ($Z = 56$), etc. isotopes. Progressively, the pattern suggests a smooth onset of deformation (cf. figure 2.7, 2.8 and 2.14). In turn, the pattern of 2_2^+ excitations can be followed: figure 4.34 shows the systematics of these states and their associated bands in the even-mass Xe isotopes. At present it is not possible to provide a structural analysis of these states in a systematic manner because there are insufficient lifetime data and consequently insufficient $B(E2)$ data.

Odd-mass nuclei in this mass region exhibit some features that have been amenable to interpretation. The most significant systematic feature is shown in figure 4.35. The data, for bands built on a state with spin-parity $11/2^-$ in the odd-La isotopes, have been interpreted as dominated by the unpaired proton occupying the $1h_{11/2}$ orbital with its spin aligned parallel to the Ba core spins. This has been given the name the 'rotation-aligned' coupling scheme. It is seen to be simply the limiting case of the coupling shown in figure 3.10 (left-hand side).

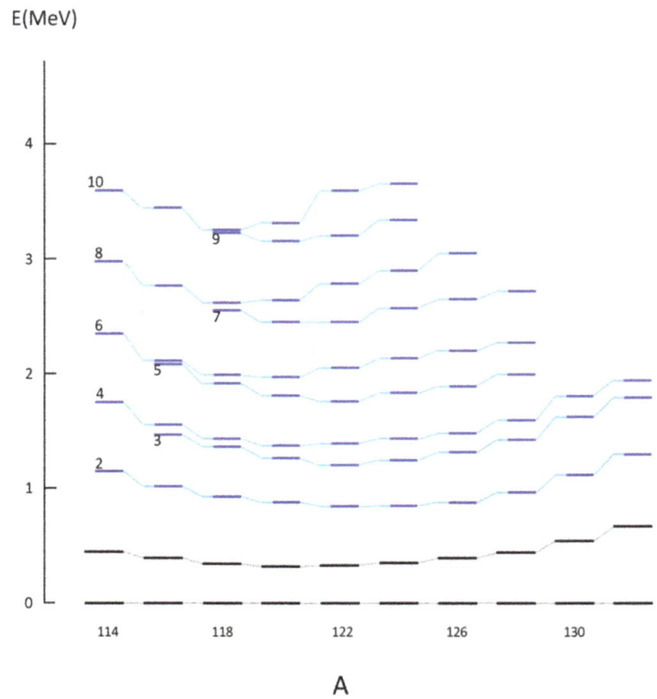

Figure 4.34. Systematics of bands built on top of 2_2^+ states in the even-Xe ($Z = 54$) isotopes. Some more details are given in the text.

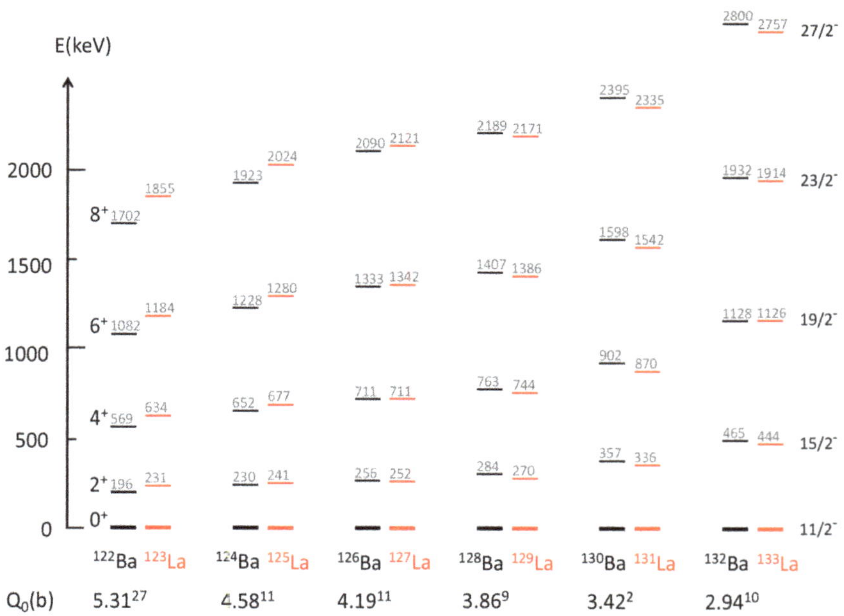

Figure 4.35. Systematics of bands bult on $11/2^-$ states in the odd-mass La ($Z = 57$) isotopes. Some more details are given in the text. This figure is an up-to-date view of these isotopes, inspired by a figure that appeared in [25].

At present, the organization of nuclear structure data for weakly deformed nuclei must be classified as a research task. Thus, we defer an in-depth analysis of such nuclei until later in the Series. In particular, we note that, in surveying the data on weakly deformed nuclei we have found a severe lack of useful systematic information. Only limited exploration has been made for such nuclei. Often, there are only data for high-spin, yrast states. We express the opinion that this situation in weakly deformed nuclei is the single greatest deficiency in advancing our understanding of nuclear structure. While there is an abundance of model fitting to limited data sets; historically this has proven, too often, to lead along paths to serious misconceptions about the structures involved.

We adopt the view that the path to a unified view of nuclear structure lies in the realm of weakly deformed nuclei. The manifestation of strongly deformed and weakly deformed, even spherical (shape coexisting) states in a single nucleus offers insight into the direct competition occurring when many nucleons self-organize. A major question is 'does shape coexistence occur in all nuclei?' Even further: 'how many different shapes are possible?'

4.5 Exercises

The relative challenge and time investment associated with the exercises is indicated from * to ***. Exercises marked with '**E**' require interactive access to figures or watching included video exercises and will only have such functionality in the e-book version.

4-1. Make systematics plots similar to those in figure 4.10 for the $N = 62$, 64, and 68 isotones, using data from ENSDF. **

4-2. Make a plot of the Pb ($Z = 82$) seniority systematics, similar to figure 4.18 and 4.19 for the closed shells at $N = 82$ and $Z = 50$. Hint: the dominant j value will be 13/2 from the neutron $1i_{13/2}$ orbital. ***

4-3. With reference to your result in exercise 4-2 and figures 4.21 and 4.28, illustrate the effects of the $3p_{1/2}$ orbital and pairing. **

4-4. Using the m scheme, deduce the possible couplings in terms of J and M
   (a) for two identical fermions with $j = 7/2$, e.g. coupling two $f_{7/2}$ neutrons. *
   (b) for three identical fermions with $j = 7/2$, and obtain the allowed spins for the $v = 3$ states. Compare your result with information in figure 4.25. **
   (c) for two non-identical fermions with $j = 7/2$, e.g. coupling an $f_{7/2}$ proton and $f_{7/2}$ neutron. **

4-5. Using the m scheme, deduce the possible couplings in terms of L and M for two identical bosons with $l = 2$. *

4-6. Using ENSDF and information in figures 4.28, 4.30; 6.14, 6.15, 6.16, classify the lowest states in ^{209}Bi, i.e. shell-model particle states, shell-model hole states, and core-excited states coupled to shell-model particle states. ***

4-7. Using the shell-model states identifiable in figure 4.28, make a catalog of the expected lowest-energy states in ^{210}Bi, i.e. using experimental energies from ^{209}Bi, ^{209}Pb. **

4-8. Make a plot of the observed energies (ENSDF) versus spin for the lowest states in ^{210}Bi with negative parity and J = 0, 2, 4, 6, 8, and separately J = 1, 3, 5, 7, 9. (Compare the even-spin pattern with figure 4.20.) **

4-9. Using information in figure 4.28, make a catalog of the expected lowest-energy excited states in ^{208}Pb, i.e. proton 1p–1h and neutron 1p–1h states. Compare your results with ENSDF for ^{208}Pb. (For the experts, take a look at [26]. Further, for the experts, take a look at [27].) **

4-10. Make a comparison of the observed energies (ENSDF) for the states in ^{42}Ca, ^{42}Sc, and ^{42}Ti below 1600keV, e.g. as a side-by-side-by-side plot. Noting your answers to exercise 4.4(a), and the fact that ^{42}Ca (Z = 20, N = 22) and ^{42}Ti (Z = 22, N = 20) are a 'mirror' pair, cf. figure 2.17, what can you conclude about the behavior of the two valence nucleons in these A = 42 isobars? ***

4-11. For those reading the e-book, explore the set of video-based exercises presented in figures 4.36–4.40. These videos are short self-contained explanations of different aspects of the nuclear structure of ^{207}Pb and

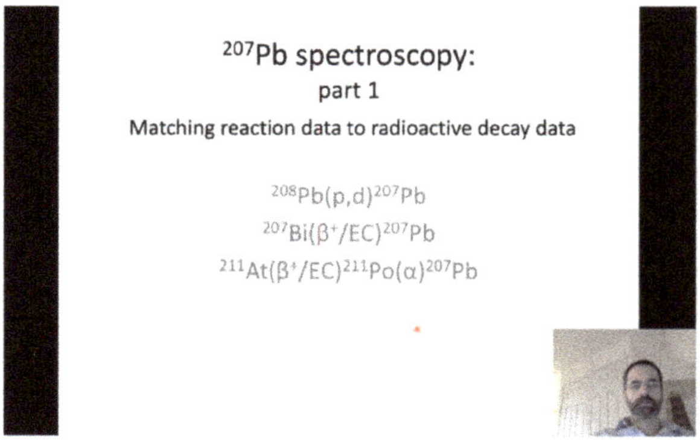

Figure 4.36. Video exercise on ^{207}Pb: Part 1. Matching reaction data to radioactive decay data. Video available at https://iopscience.iop.org/book/978-0-7503-2674-2.

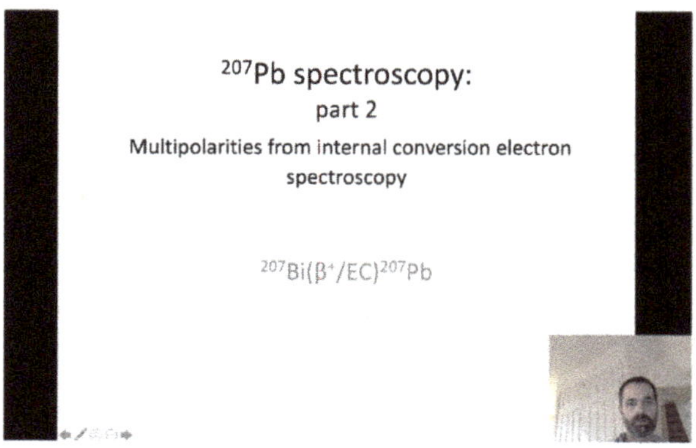

Figure 4.37. Video exercise on ^{207}Pb: Part 2. Multipolarities from internal conversion electron spectroscopy. Video available at https://iopscience.iop.org/book/978-0-7503-2674-2.

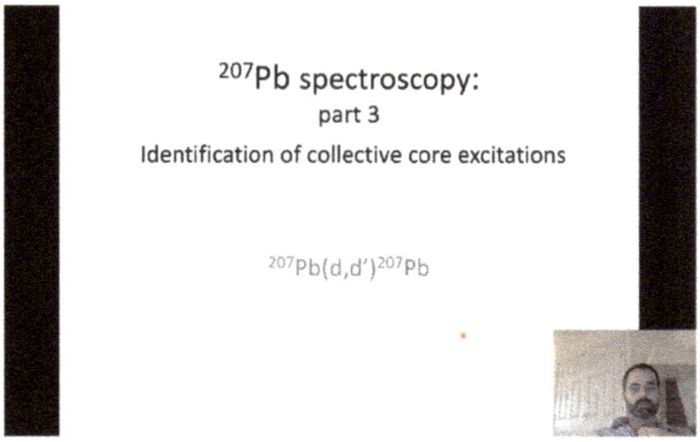

Figure 4.38. Video exercise on ^{207}Pb: Part 3. Identification of collective core excitations. Video available at https://iopscience.iop.org/book/978-0-7503-2674-2.

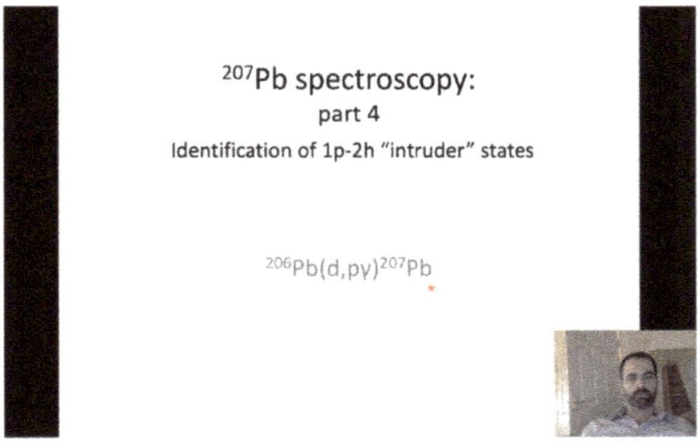

Figure 4.39. Video exercise on ^{207}Pb: Part 4. Identification of 1p–2h 'intruder' states.Video available at https://iopscience.iop.org/book/978-0-7503-2674-2.

Figure 4.40. Video exercise on ^{207}Pb: Part 5. A comprehensive spectroscopic view. Video available at https://iopscience.iop.org/book/978-0-7503-2674-2.

contain short exercises intended to encourage familiarity with identifying states adjoining the doubly magic nucleus, ^{208}Pb, and to draw on many of the concepts and equations presented in this chapter. The powerpoints presented in the short videos can also be accessed here: [https://iopscience.iop.org/book/978-0-7503-2674-2]. *** **E**

References

[1] Heyde K and Wood J L 2016 Nuclear shapes: from earliest ideas to multiple shape coexisting structures *Phys. Scr.* **91** 083008

[2] Heyde K, Van Isacker P, Waroquier M, Wood J L and Meyer R A 1983 Coexistence in odd-mass nuclei *Phys. Rep.* **102** 291

[3] Wood J L, Heyde K, Nazarewicz W, Huyse M and Van Duppen P 1992 Coexistence in even-mass nuclei *Phys. Repts.* **215** 101

[4] Wood J L, Zganjar E F, De Coster C and Heyde K 1999 Electric monopole transitions from low energy excitations in nuclei *Nucl. Phys.* A **651** 323

[5] Heyde K and Wood J L 2011 Shape coexistence in atomic nuclei *Rev. Mod. Phys.* **83** 1467

[6] Wood J L and Heyde K 2016 A focus on shape coexistence in nuclei *J. Phys.* G **43** 020402

[7] Rowe D J and Wood J L 2010 *Fundamentals of Nuclear Models: Foundational Models* (Singapore: World Scientific)

[8] Pore J L *et al.* 2017 Study of the β^- decay of ^{116m1}In: A new interpretation of low-lying 0^+ states in ^{116}Sn *Eur. Phys. J.* A **53** 27

[9] Kantele J *et al.* 1979 Absolute *E*0 and *E*2 transition rates and collective states in ^{116}Sn *Z. Phys.* A **289** 157

[10] Fielding H W *et al.* 1977 0^+ states observed in Cd and Sn nuclei with the (^{3}He,n) reaction *Nucl. Phys.* A **281** 389

[11] Shoup R, Fox J D and Vourvopoulos G 1969 Proton particle-hole states in ^{116}Sn from the ^{115}In(^{3}He,d)^{116}Sn reaction *Nucl. Phys.* A **135** 689

[12] Wimmer K *et al.* 2010 Discovery of the shape coexisting 0^+ state in ^{32}Mg by a two neutron transfer reaction *Phys. Rev. Lett.* **105** 252501

[13] Rotaru F *et al.* 2012 Unveiling the intruder deformed 0_2^+ state in ^{34}Si *Phys. Rev. Lett.* **109** 092503

[14] Hoyle F 1954 Helium fusion in stars *Astrophys. J. Suppl. Ser.* **1** 121

[15] Kibédi T *et al.* 2020 Radiative width of the Hoyle state from γ-ray spectroscopy *Phys. Rev. Lett.* **125** 182701

[16] Eriksen T K *et al.* 2020 Improved precision on the experimental *E*0 decay branching ratio of the Hoyle state *Phys. Rev.* C **102** 024320

[17] Heyde K and Wood J L 2019 *Quantum Mechanics for Nuclear Structure, Volume 1: A primer* (Bristol: IOP Publishing) https://iopscience.iop.org/book/978-0-7503-2179-2

[18] Marín-Lámbarri D J, Bijker R, Freer M, Gai M, Kokalova T, Parker D J and Wheldon C 2014 Evidence for triangular D_{3h} symmetry in ^{12}C *Phys. Rev. Lett.* **113** 012502

[19] Iskra L *et al.* 2014 Higher-seniority excitations in even neutron-rich Sn isotopes *Phys. Rev.* C **89** 044324

[20] Lozeva R L *et al.* 2008 New sub-μs isomers in ^{125,127,129}Sn and isomer systematics of $^{124\text{-}130}$Sn *Phys. Rev.* C **77** 064313

[21] Schiffer J P and True W W 1976 The effective interaction between nucleons deduced from nuclear spectra *Rev. Mod. Phys.* **48** 191

[22] Taniuchi R *et al.* 2019 ^{78}Ni revealed as a doubly-magic stronghold against nuclear deformation *Nature* **569** 53
[23] Morinaga H 1956 Interpretation of some of the excited states of 4n self-conjugate nuclei *Phys. Rev.* **101** 254
[24] Crawford H L *et al.* 2017 Unexpected distribution of $\nu f_{7/2}$ strength in ^{49}Ca *Phys. Rev.* C **95** 064317
[25] Stephens F S, Diamond R M, Leigh J R, Kammuri T and Nakai K 1972 Decoupled yrast states in odd-mass nuclei *Phys. Rev. Lett.* **29** 438
[26] Heusler A *et al.* 2016 Complete identifiction of states in ^{208}Pb below $E_x = 6.2$ MeV *Phys. Rev.* C **93** 054321
[27] Broda R *et al.* 2017 Doubly magic ^{208}Pb: High-spin states, isomers, and *E*3 collectivity in the yrast decay *Phys. Rev.* C **95** 064308

Chapter 5

Simple models of nuclear structure

The basic models that are used to organize data are introduced in the simplest terms. The shell model is outlined with a look at its shell and subshell structure. The Nilsson model is described with some details of the use of asymptotic quantum numbers and basis states. The essential content of pairing models is sketched. The rotor model is developed using an analogy with rotation of diatomic molecules. The types of band mixing that occur in terms of the K quantum number are noted. Some details of Coriolis band mixing are given. Some basic equations describing moments of inertia for extended deformed bodies are derived. The Bohr model is briefly outlined.

Concepts: independent particle, potential well, well-flattening, spin–orbit coupling, cylindrical potential, volume conservation, Ω and *K* quantum numbers, Cooper pair, moment of inertia, Coriolis term, adiabatic approximation.

Nuclear models serve a number of roles in the study of nuclear structure. First, they play a critical role in the organization of data. Second, they provide important 'shadows' of the underlying quantum mechanics that leads to a description of the many-body behavior of nuclei. Any quantum mechanical many-body theory of the nucleus must, ideally, possess the simple models as limiting cases. Herein, we sketch a few of the basic quantum mechanical details of the simple models that have been used to organize the data in chapters 1–4.

5.1 Independent-particle models for nuclear structure

Independent-particle models play a number of key roles in the study of nuclear structure. Of primary interest herein is their role in the organization of data. They also play a fundamental role as a quantum mechanical *basis*. This second role is a key component of many-body quantum mechanics: we employ descriptions of many-body systems using many copies of one-body configurations. These are formulated as sums of single-particle energies plus terms describing the interactions between the particles, and direct products of the configurations. The configurations

doi:10.1088/978-0-7503-2674-2ch5

are taken from independent-particle models; for example, the simple harmonic oscillator model provides a widely used basis.

Independent-particle models of nuclei appear in two extreme forms: the (spherical) shell model and the (deformed) Nilsson model. It is always necessary to modify these models by introducing correlations. Pairing correlations are present in all nuclei, minimally so in doubly closed shell plus or minus a single nucleon, e.g. $^{207,\,209}$Pb, ^{207}Tl, ^{209}Bi, cf. figure 4.28. The challenge is the description of nuclei that lie between the extremes of spherical shapes and strongly deformed shapes. Many nuclei lie between these extremes: they are often termed 'transitional nuclei'. It is in these nuclei that one finds an active research frontier; and a lack of consensus with respect to interpretation—especially with regard to energy contributions from independent-particle behavior as contrasted with contributions from correlations. A key unresolved feature of these transitional nuclei is the presence of deformation—specifically, what is the smallest deformation that can be deduced from data? We provide basic details of the spherical shell model and the Nilsson model.

5.1.1 The shell model

The spherical shell model is the longest-standing quantum mechanical model of nuclear structure. As such, it is widely familiar to most physicists; and, as such, it imposes a bias on the thinking of nuclear physicists.

The Hamiltonian for the shell model contains three essential terms, expressed as

$$H = \frac{p^2}{2m} + V(r) + C\,\mathbf{l}\cdot\mathbf{s}, \tag{5.1}$$

where $p^2/2m$ is a kinetic energy term, $V(r)$ is a spherically symmetric potential energy term, and $C\,\mathbf{l}\cdot\mathbf{s}$ is the so-called spin–orbit coupling term. Empirically, it is determined[1] that $C < 0$. It was the recognition of the need for a large spin–orbit coupling interaction in nuclei (as contrasted with atoms) that was the key step taken in 1949 by Maria Goeppert Mayer and by Hans Jensen and co-workers that launched the shell model as the first quantum mechanical model of nuclear structure.

There are a variety of choices made for $V(r)$. Figure 5.1 illustrates the three most popular choices. Each choice has pros and cons. The most realistic choice is the so-called Woods–Saxon potential. This matches the density distribution of nucleons in the nucleus, viz.

$$V(r) \sim -\rho(r), \tag{5.2}$$

where $\rho(r)$ is the density distribution of nuclear mass within the nucleus, cf. figure 1.3. Note, it is assumed that $\rho_{\text{mass}} \sim \rho_{\text{charge}}$. Recognizing that the nucleon–nucleon interaction is short-ranged, this is the best model potential; however, it cannot be solved in 'closed form', i.e. an energy eigenvalue equation is not available and the energy eigenstates must be expanded in a basis (such as a harmonic

[1] The attractive nature of the spin–orbit interaction has recently been reported in a derivation by Chris Halcrow and Derek Harland [1].

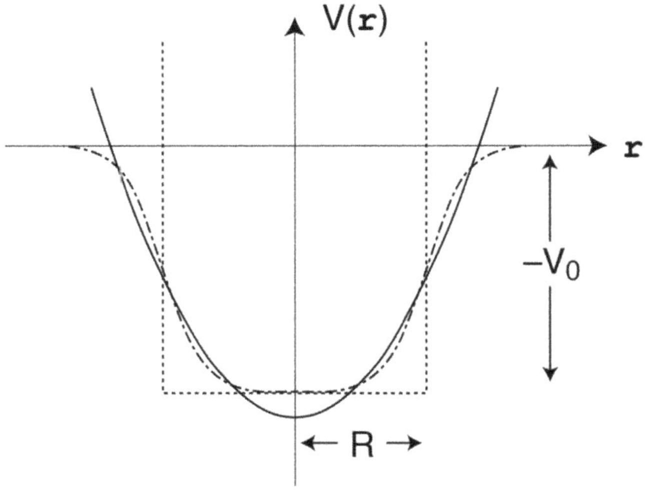

Figure 5.1. The three commonly used radial potentials for independent-particle central-force motion in the atomic nucleus. The Wood–Saxon potential is depicted by the dotted-dashed line, the infinite-square well potential by the dashed line, and the harmonic oscillator potential by the solid line. The depth of the Woods–Saxon potential is given as $-V_0$. The radius of the infinite-square well potential is given as R. The depiction is for 'positive and negative values of r' to convey the view of the entire nucleus, but r only takes positive values. See the text for further details. Reproduced from [3], copyright (2010) World Scientific Publishing Company.

oscillator basis). The first simple choice that can be solved in closed form is the infinite square well potential, viz.

$$V(r) = \text{const.},\ r < R, \tag{5.3}$$

where R is the root-mean squared radius of the nucleus, given by $R = 1.2A^{1/3}$ fm. The energy eigenvalues are depicted in figure 5.2. The energy eigenfunctions and the procedure for obtaining the energy eigenvalues are given in [2], chapter 12. The second simple choice that can be solved in closed form is the harmonic oscillator potential, viz.

$$V(r) = \frac{1}{2}m\omega^2 r^2, \tag{5.4}$$

where m is the mass (bare mass) of the nucleon (recall $m_n \sim m_p$ within 1.3 MeV), and the energy scale factor is given by

$$\hbar\omega = 41A^{-1/3}\ \text{MeV}. \tag{5.5}$$

The energy eigenvalues are depicted in figure 5.3. Again, details are given in [2], chapter 12. A key modification of the oscillator potential is essential: it requires a so-called 'well-flattening' term that can be modeled by adding a term Dl^2 to the Hamiltonian, with $D < 0$. This term effectively lowers energies of higher angular momentum eigenstates, which is consistent with a 'flatter' potential favoring such states, cf. the energy patterns in figures 5.2 and 5.3. This is also illustrated in [2], figure 12.1. The infinite square well and the harmonic oscillator potentials

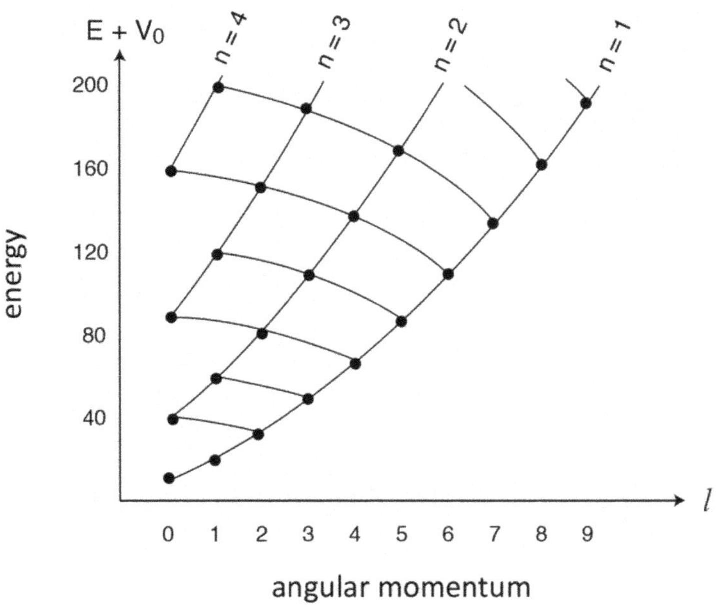

Figure 5.2. Energy eigenstates for an independent-particle Hamiltonian with a three-dimensional isotropic infinite square-well potential. Note that for each angular momentum value, $l = 0, 1, 2, 3, \ldots$ there is an energy degeneracy given by $m = l, l - 1, l - 2, l - 3, \ldots, -l + 1, -l$; further there is a spin degeneracy, ×2. See the text for further details. Reproduced from [3], copyright (2010) World Scientific Publishing Company.

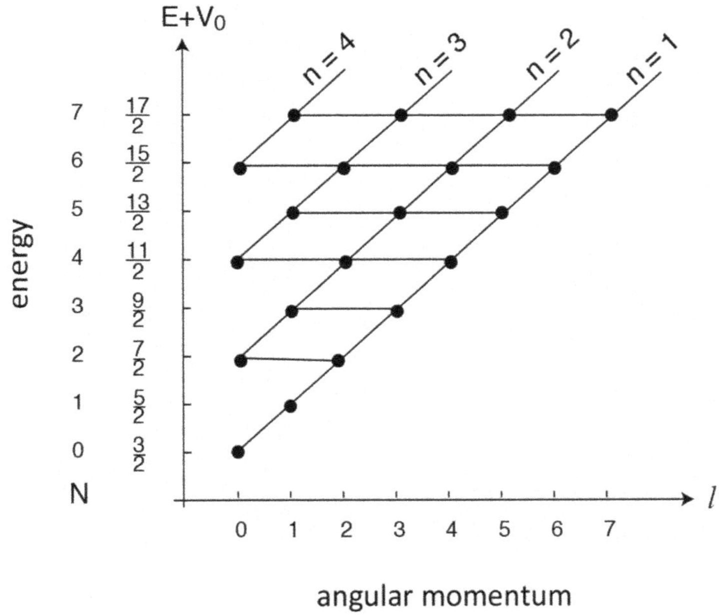

Figure 5.3. Energy eigenstates for an independent-particle Hamiltonian with a three-dimensional isotropic harmonic oscillator potential. See the caption to figure 5.2 for the details of energy degeneracy with respect to angular momentum and spin. See the text for further details. Reproduced from [3], copyright (2010) World Scientific Publishing Company.

are infinitely deep and so they do not offer descriptions for weakly bound states in nuclei.

A qualitative view of the shell model energy level diagram is depicted in figure 5.4. The most important aspect of this diagram is that it reveals an energy level *ordering* as a function of angular momentum and total spin. The exact ordering within a

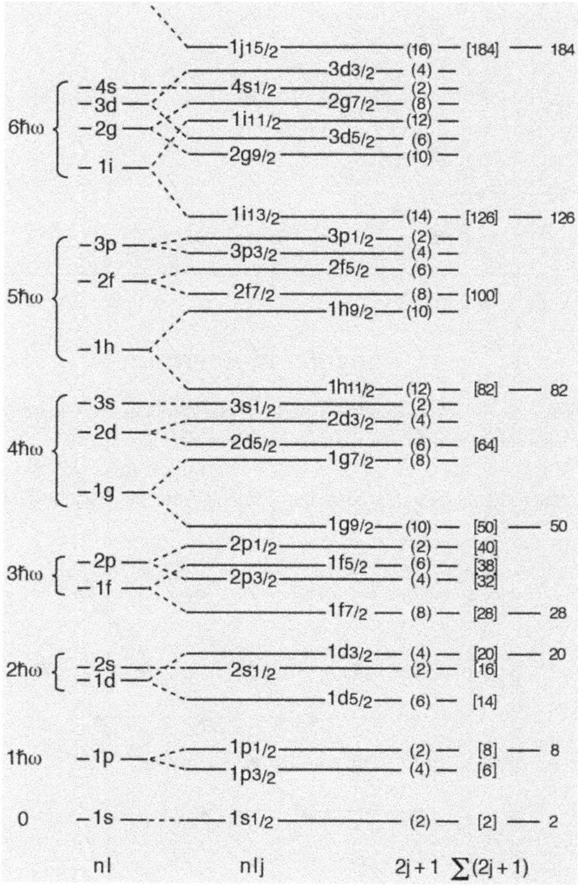

Figure 5.4. A shell model energy level diagram. The pattern of levels is only qualitative: it adopts an isotropic harmonic oscillator central potential; but it does recognize the effect of 'well-flattening', cf. figure 5.5, and it incorporates a large, empirically determined spin–orbit splitting. States are labeled with quantum numbers dictated by spin–orbit coupling, i.e. $j = l + s$. Parity is dictated as $\pi = (-1)^l$. Degeneracies of the j subshells, where $m_j = +j, +j - 1, \ldots, -j$, i.e. $(2j + 1)$ states per subshell, are indicated. The cumulative filling numbers of nucleons, separately for protons and neutrons, are given in square brackets. We emphasize, nuclei that might be considered to exhibit states manifesting shell model character must be handled with caution when making interpretations because of the presence of correlations: this applies especially to where observed energies can only be equated with this diagram in carefully selected nuclei, such as presented in figures 4.28, 5.5, and 5.6. Energies of states labeled with shell model quantum numbers, such as shown in figures 4.13, 4.14, 4.25, 4.30, 4.31, 5.13, and 5.15, must be treated with considerable caution when making inferences regarding energies of states that appear to match shell model expectations. Reproduced from [3], copyright (2010) World Scientific Publishing Company.

given energy shell is subject to details, some of which are discussed below. However, always caution is needed when comparing with data because correlations can cause energy shifts that are not always obvious or anticipated. A useful view of the shell model, as applied to nuclei adjacent to doubly closed shells is depicted in figure 5.5. This shows the result of removing the spin–orbit coupling energy, scaled by an empirically determined value of $C = -0.5$ MeV, to reveal the energies of the various l-value eigenstates in a given oscillator shell. The spin–orbit coupling energies are given by

$$E_{ls} = +\frac{1}{2}Cl\hbar^2, j = l + \frac{1}{2}, \tag{5.6}$$

and by

$$E_{ls} = -\frac{1}{2}C(l + 1)\hbar^2, j = l - \frac{1}{2} \tag{5.7}$$

where details of their derivation are given in [4], chapter 2, section 2.3.

A leading question, with respect to the shell model, is how does the model fail? It is widely recognized that correlations must be introduced in nuclei, in order to achieve a comparison between data and the shell model. However, in odd-mass nuclei adjacent to doubly closed shells, some direct comparisons are possible. Figure 5.6(a) and (b) illustrates an important failure of the shell model. Evidently,

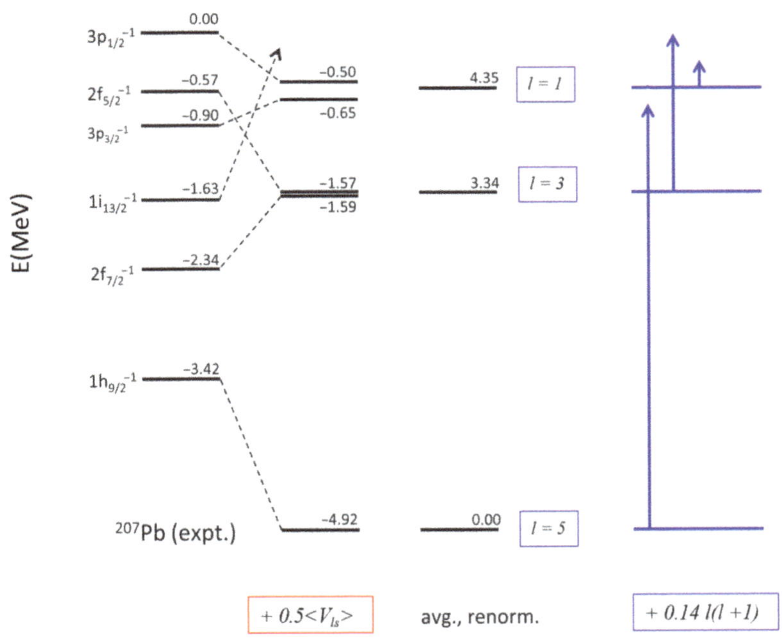

Figure 5.5. An illustration of the need to have a term in the mean-field Hamiltonian that separates shell model orbitals with different angular momenta in an ordered pattern. The strength of the spin–orbit potential is taken from global empirical data. The strength of the $l(l + 1)$ term is chosen to minimize the energy splitting shown here. The excitation pattern shown favors a 'flat-bottomed' potential, cf. figures 5.2 and 5.3.

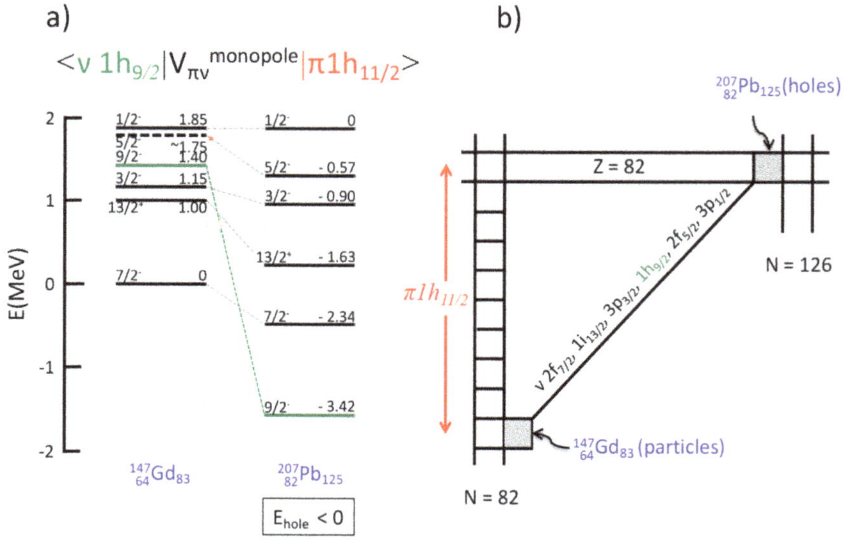

Figure 5.6. 'Migration' of specific shell model orbitals, as inferred from states observed in ^{147}Gd and ^{207}Pb. The energy of the state in ^{147}Gd with spin-parity $5/2^-$, viz. ~1.75 MeV, is from systematics in the $N = 83$ isotones. Note that the energies in ^{207}Pb are expressed as negative quantities, i.e. as neutron holes: this can be viewed as 'looking down into the potential well from above the closed shell'. Further details are discussed in the text. The shift in energy of the $9/2^-$ state, interpreted as the $\nu 1h_{9/2}$ shell model configuration, is attributed to filling of the $\pi(3s_{1/2} + 2d_{3/2} + 1h_{11/2})$ subshells by 18 protons. Notably, a short-ranged attractive interaction between the unpaired neutron in the $\nu 1h_{9/2}$ orbital and the protons occupying the $\pi 1h_{11/2}$ orbital can explain this. The short-ranged interaction is expressed as $\langle \nu 1h_{9/2}| V_{\pi\nu}^{\text{mono}} |\pi 1h_{11/2}\rangle$, where 'mono' indicates that it is a so-called 'monopole' interaction, i.e. it only acts between states with the same angular momentum (h orbitals have $l = 5$).

the filling of the proton subshells influences the neutron single-particle energies. This is attributed to an enhanced interaction between neutrons in the $1h_{9/2}$ orbital and protons filling the $1h_{11/2}$ orbital, based on simple orbital overlap considerations and a short-ranged interaction between the protons and neutrons.

Key spectroscopic probes of independent-particle behavior in nuclei are one-nucleon transfer processes and one-nucleon addition/knockout processes. One never observes a true single-particle or single-hole strength. Experimental data can be reduced to so-called spectroscopic factors, which are a measure of independent particle or hole strength. Breakdown of the shell model is manifested in 'fragmentation' of strength and in the need to sum spectroscopic strength for a given total spin (and parity) value to arrive at energy centroids for the configuration. An example is illustrated in figure 5.7. Quantification of this approach in assessing the presence of independent-particle and independent-hole strength is a complex process that lacks a definitive treatment; and it is overshadowed by severe warnings regarding the validity of the concept of an independent particle or hole in a complex many-body system, already at the most fundamental level. Delving further into this issue involves what we would term 'advanced' nuclear structure and lies beyond the scope of the present treatment.

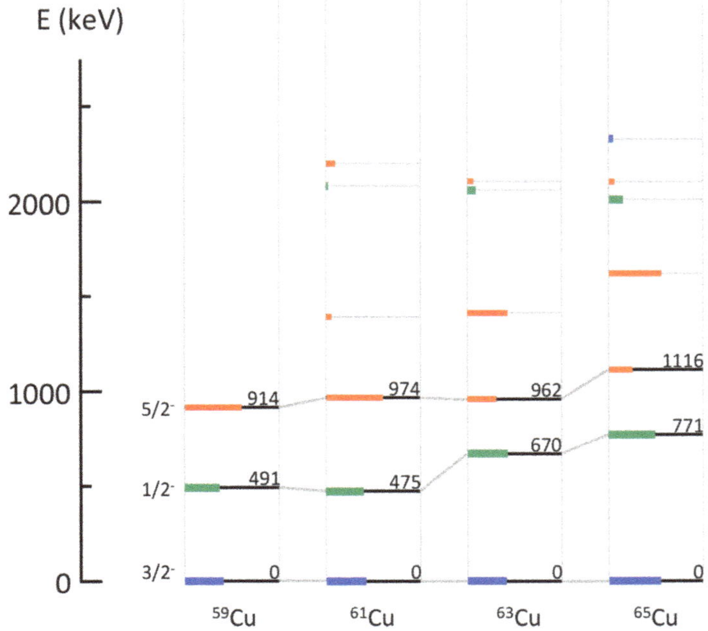

Figure 5.7. The lowest three states in $^{59-65}$ Cu and states for which there is significant population in the one-proton transfer reaction (^{3}He,d). The lengths of the colored bars are directly proportional to the strength of population of the states in the transfer reaction. The ground-states can be naively interpreted as the shell model configuration $2p_{3/2}$; similarly, the first excited states would appear to be the configuration $2p_{1/2}$. However, the assumption that these are spherical shell model states breaks down for the second excited state where the one-proton transfer strength is severely fragmented. The color coding indicates the l values of the transferred nucleon—blue ($l = 1$ with $j = l + 1/2$), green ($l = 1$ with $j = l - 1/2$), red ($l = 3$ with $j = l - 1/2$). The direct interpretation is that the l quantum number is not a good quantum number in these nuclei, i.e. they are deformed. Further discussion of such issues must await a more advanced level of treatment. The data are taken from [5].

5.1.2 The Nilsson model

The Nilsson model followed shortly after the spherical shell model. It is simply a deformation of the potential term in the shell model, usually taken in the harmonic oscillator form, maintaining an axial (cylindrical) symmetry[2]. The Hamiltonian is expressed as

$$H = \frac{p^2}{2m} + \frac{m}{2}\left[\omega_{xy}^2(x^2 + y^2) + \omega_z^2 z^2\right] + Dl^2 + C\,\mathbf{l}\cdot\mathbf{s}. \tag{5.8}$$

[2] At a more fundamental level, the deformation of the shell-model spherical mean field can be viewed as an example of the so-called Jahn–Teller effect. When (near) degenerate configurations exist at the energy of the ground state of a system, mixing occurs in such a manner that 'distortion' of the system occurs. In nuclei, the shell model subshells (labeled by the quantum numbers j and l) are sufficiently close in energy that this produces deformation of the nucleus. Thus, a signature of the onset of deformation in weakly deformed nuclei is the 'fragmentation' of j values, cf. figure 5.7. For the advanced Reader, we note [6], wherein it is suggested that the origin of the mixing is the so-called nuclear giant quadrupole resonance. Such issues will be addressed later in the Series.

The simplest solution for the energy eigenvalues of the Hamiltonian, equation (5.8), the 'asymptotic' solution, corresponds to the extreme cylindrically deformed limit where the harmonic oscillator part leads to energy eigenvalues, viz.

$$E(N, n_z) = (N - n_z + 1)\hbar\omega_{xy} + \left(n_z + \frac{1}{2}\right)\hbar\omega_z. \tag{5.9}$$

If this expression were the end of the model story, one sees that by making the elongation axis become infinitely 'stretched' one allows many particles in such a potential to take on a total energy that could approach zero. Nuclei do not do this: a constraint must be imposed such that the 'shape' of the potential matches the shape of the nucleus. 'Shape' here equates to the quadrupole moment of the nucleus. The connection between the observable (quadrupole moment) and the model parameter that describes the deformation is contained in the equations defining an 'equipotential' surface, viz.

$$\omega_{xy}^2 R_x^2 = \omega_{xy}^2 R_y^2 = \omega_z^2 R_z^2, \tag{5.10}$$

via

$$\omega_{xy} = \omega_0 \exp\left(\frac{\epsilon}{3}\right), \; \omega_z = \omega_0 \exp\left(-\frac{2\epsilon}{3}\right), \tag{5.11}$$

from which a constant volume, proportional to $R_x R_y R_z$, is seen to be consistent with the 'deformation' parameter ϵ. This parameter is connected to a quadrupole moment via the relationship

$$V = \frac{m\omega_0 r^2}{2} - \frac{4}{3}\sqrt{\frac{\pi}{5}}\epsilon m\omega_0 r^2 Y_{20}(\theta, \phi), \tag{5.12}$$

where $Y_{20}(\theta, \phi)$ is the quadrupole spherical harmonic with axial symmetry, and recall

$$r^2 Y_{20}(\theta, \phi) = \sqrt{\frac{5}{16\pi}}(2z^2 - x^2 - y^2). \tag{5.13}$$

The eigenvalues of $V_{\text{def}} := -\frac{4}{3}\sqrt{\frac{\pi}{5}}\epsilon m\omega_0 r^2 Y_{20}(\theta, \phi)$ are given by

$$E_{\text{def}} = \frac{2}{3}\epsilon\left(N + \frac{3}{2}\right)\hbar\omega_0 \frac{[3\Omega^2 - j(j+1)]}{4j(j+1)}, \tag{5.14}$$

where $\Omega := j_z$. For a single j subshell, the energies E_{def} are depicted in figure 3.7(a). A useful semi-classical view of E_{def} is given in figure 3.7(c) in terms of orbital alignments within the deformed mean field of the nucleus.

Equation (5.14) expresses the diagonal matrix elements of V_{def}. This term also has off-diagonal matrix elements: these cause mixing of different j values but preserve Ω as a good quantum number for the model. Because the deformed potential is reflection symmetric, parity is also conserved. In consequence the mixing matrix that must be diagonalized involves a subspace of Hilbert space that is characterized simply by configurations with the same Ω, π values, cf. figure 3.7. Energy eigenstates are expressed in the spherical basis as

$$|\Omega, \pi\rangle = \sum_{jl} C_{jl}|Njl\Omega\pi\rangle. \tag{5.15}$$

The influence of $V_{\rm def}$ can extend to mixing $|\Omega\pi\rangle$ configurations differing in N by 2, so-called $\Delta N = 2$ mixing. This only occurs in very isolated cases; but then the summation in equation (5.15) must extend also over N values.

Examples of a so-called Nilsson diagram are depicted in figures 5.8 and 5.9. In using a Nilsson diagram, one needs to know the number of nucleons, e.g. for Lu, $Z = 71$, and an estimate of the deformation parameter, e.g. from a quadrupole moment of the nucleus of interest or of a neighboring nucleus. The diagrams, as depicted are only approximate with respect to level order for a given deformation,

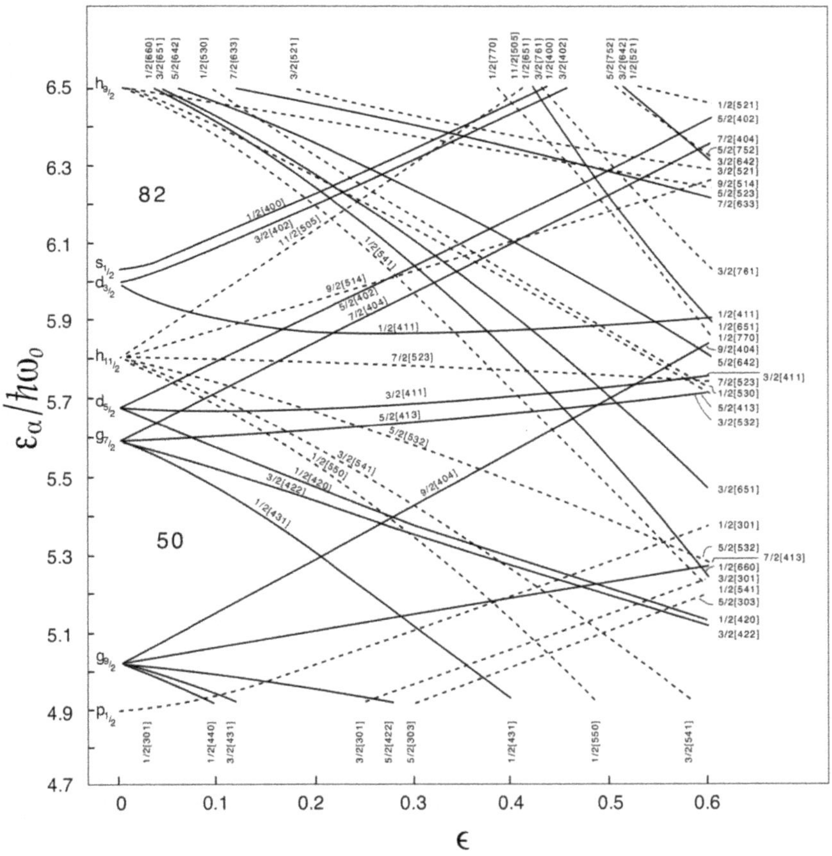

Figure 5.8. Nilsson diagram for odd-proton nuclei in the rare-earth region. All Nilsson states are two-fold degenerate. The states are labeled using the asymptotic quantum numbers, $\Omega[Nn_z\Lambda]$ (see text). A useful relationship between the asymptotic quantum numbers and the spherical shell model quantum numbers is $n_z + \Lambda = l$. Solid lines are positive-parity states, dashed lines are negative-parity states. Reproduced from [3], copyright (2010) World Scientific Publishing Company.

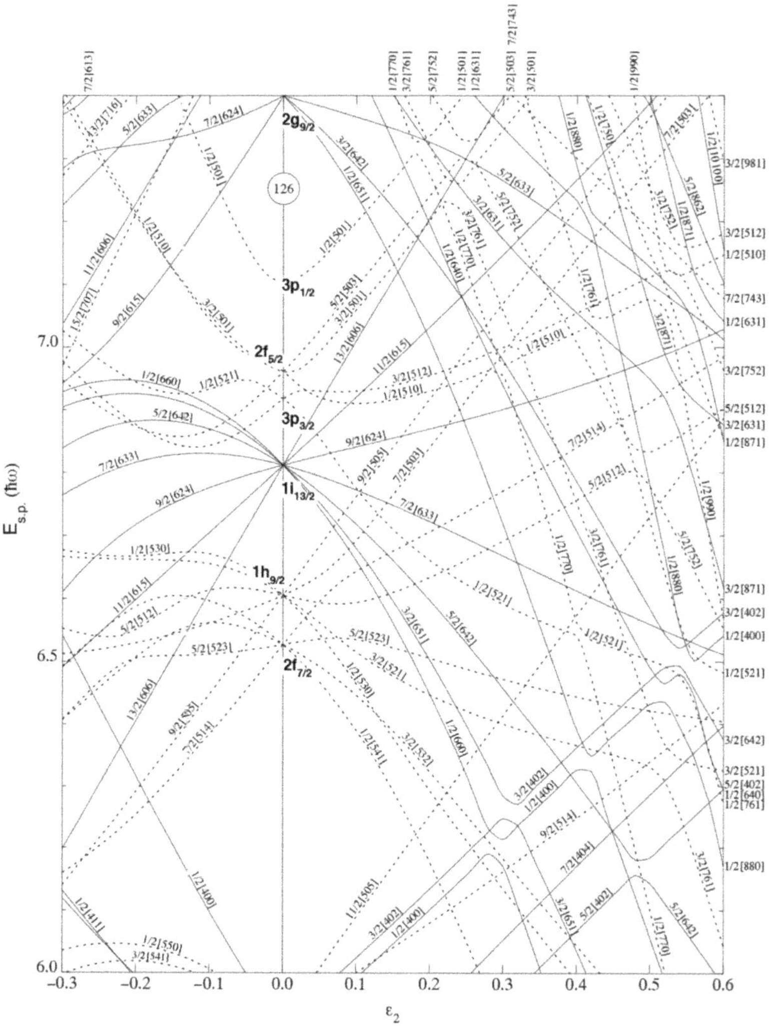

Figure 5.9. Nilsson diagram for odd-neutron nuclei in the rare-earth region (see caption to figure 5.8 for details). The figure is reproduced with permission of John Wiley & Sons, from [7]; permission conveyed through Copyright Clearance Center, Inc.

and are never quantitative. The primary reason for lack of quantitative 'predictions' is the ever-present role of correlations, particularly pairing correlations. Details of pairing correlations follow in the next section.

The asymptotic basis leads to only diagonal energies for V_{def}. It is the 'spherical' terms, $C\,\mathbf{l}\cdot\mathbf{s}$ and Dl^2 in the Hamiltonian that mix this basis. For the interested Reader, discussion of solutions to the Nilsson Hamiltonian in the asymptotic basis are given in, e.g. [8]. The main value of the asymptotic basis is as a set of approximate labeling quantum numbers. Because the spin–orbit term is a mixing term, the asymptotic basis adopts $\Lambda := l_z$ and $\Sigma := s_z$ as approximately good quantum numbers, viz. $|Nn_z\Lambda\Sigma\rangle\Omega^{\pi}$.

The solutions to the Nilsson model Hamiltonian are independent of the basis used. The asymptotic quantum numbers are the popular and universally used labels of configurations. The spherical basis is the universally preferred basis for diagonalization. The spherical basis plays a crucial role in the interpretation of one-nucleon transfer reactions. The reason for this is that the reaction takes place in the laboratory frame in which the spherical basis provides good quantum numbers. In practical terms, the C_{jl} coefficients in equation (5.15) take on physical significance in the definition of one-nucleon transfer strengths leading to final states in a deformed odd-mass nucleus, so-called 'spectroscopic factors'.

A more specialized topic is the incorporation of deformations beyond Y_{20}, e.g. Y_{22}, Y_{40}, $Y_{30}, \cdots$. These will be mentioned in the section in this chapter on models for collective rotations and vibrations. A semi-classical view of the effect of Y_{40} deformation on Nilsson states is shown in figure 5.10.

Deformation in nuclei is quantified by measured moments and is connected to model deformation parameters. The model deformation parameters appear as two basic types. The first is in model descriptions of the 'shape' of the nuclear surface, treated as that of a body with constant density and a sharp surface, i.e. not as the observed diffuse density distribution shown in figure 1.3. The second is in model descriptions of the 'shape' of the mean field for independent-particle degrees of freedom. Generally, the surface shape parameters appear as β_2, β_3, β_4, $\cdots$ and the potential shape parameters appear as ϵ_2, ϵ_3, ϵ_4, $\cdots$; with the abbreviations $\beta_2 = \beta$ and $\epsilon_2 = \epsilon$ often made when there is no risk of confusion. Where important, we provide relationships for these quantities, e.g. equations (3.26) and (5.11). These relationships possess higher-order corrections (see the following equation (5.36)). The relationship between β and ϵ (β_2 and ϵ_2) is approximately $\epsilon_2 \sim 0.95\beta_2$. These details will be addressed later in the Series.

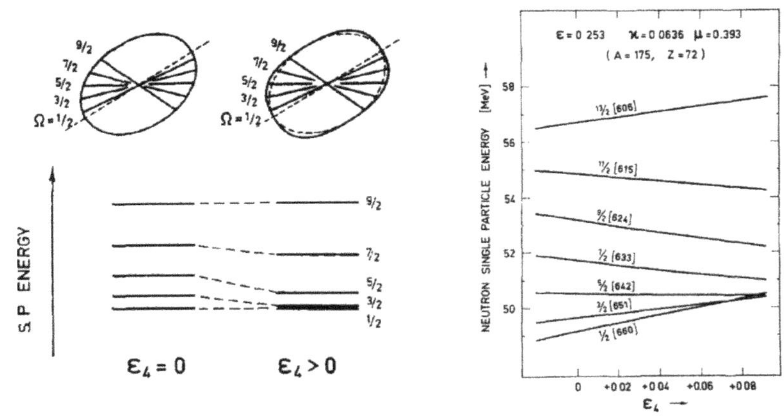

Figure 5.10. A schematic view of the influence of positive hexadecapole deformation on a single j orbital in a prolate-deformed Nilsson model potential. The figure can be viewed using 'deformation alignment', as depicted on the right-hand side of figure 3.7. Note that the left-hand energy-level diagram is for relative energies for a $j = 9/2$ orbital and the right-hand energy-level diagram is for absolute energies and is for a $j = 13/2$ orbital. Reprinted from [9], copyright (1973) with permission from Elsevier.

5.2 Pairing models for nuclear structure

'Pairing' is present in all many-nucleon systems. In consequence, models for the description of pairing in nuclei are very well established, to a high level of detail. For present purposes only a few basic features are needed and these are formulated using a simple schematic model.

We handle the essentials of pairing using the model depicted in figure 5.11. This presents a model Hamiltonian for a two-level system with two particles. The two levels are defined as two j subshells, j_1 and j_2. From the data presented in section 4.2, specifically figures 4.18 and 4.19, we make the extreme approximation that one effect of the 'pairing' force is to favor only the $J = 0$ spin coupling over the other, $J = 2, 4, 6, \ldots$ spin couplings. Thus, for the two subshells, states $|j_1^2; J = 0\rangle$ and $|j_2^2; J = 0\rangle$ are lowered in energy by $-G$, relative to the states $|j_1^2; J \neq 0\rangle, |j_2^2; J \neq 0\rangle$. We then focus on the two-dimensional $J = 0$ subspace spanned by $|j_1^2; J = 0\rangle$ and $|j_2^2; J = 0\rangle$. Naively, this subspace might be considered to be the result that we are seeking (this is a commonly held view). But there are no correlations yet. Correlations occur because of the presence of off-diagonal matrix elements. These are handled in the following.

For the two particles in the two subshells, figure 5.11, to reduce the problem to its essentials, we make the two subshells degenerate in energy, i.e. $E_1 = E_2 = E$; further, we introduce off-diagonal matrix elements, $-G$, i.e. equal to the diagonal matrix elements. (This retains the essential physics and makes the solution easy to obtain.) The Hamiltonian, in matrix form is

$$H = \begin{pmatrix} E - G & -G \\ -G & E - G \end{pmatrix} = E\begin{pmatrix} 1 & 0 \\ 0 & 1 \end{pmatrix} - G\begin{pmatrix} 1 & 1 \\ 1 & 1 \end{pmatrix}. \tag{5.16}$$

To diagonalize this Hamiltonian, we must find the solution to the matrix equation

$$\begin{pmatrix} 1 & 1 \\ 1 & 1 \end{pmatrix}\begin{pmatrix} \alpha \\ \beta \end{pmatrix} = \lambda\begin{pmatrix} \alpha \\ \beta \end{pmatrix}, \tag{5.17}$$

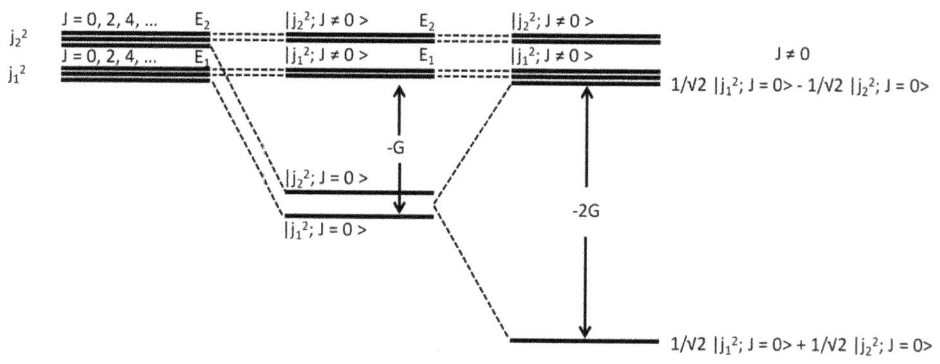

Figure 5.11. A simple model of two-state mixing that illustrates the essential physics of pairing correlations in nuclei. Details are discussed in the text.

this yields the secular equation

$$\begin{vmatrix} 1-\lambda & 1 \\ 1 & 1-\lambda \end{vmatrix} = 0, \tag{5.18}$$

i.e.

$$(1-\lambda)^2 - 1 = 0, \tag{5.19}$$

from which the solutions, $\lambda = 0$ and 2 are obtained. This result is depicted in figure 5.11: a single state has gained a binding energy of $-2G$, it is separated from all of the other states by the so-called 'pairing gap'. Such a manifestation of pairing in nuclei is presented in figure 4.21. This is a realization of the pairing model presented in [4], chapter 4, section 4.5.1, and depicted therein as figure 4.3.

The energy eigenvectors for the above two-level pairing problem are of the form

$$|\lambda = -2G\rangle = \frac{1}{\sqrt{2}}|j_1^2; J=0\rangle + \frac{1}{\sqrt{2}}|j_2^2; J=0\rangle \tag{5.20}$$

and

$$|\lambda = 0\rangle = \frac{1}{\sqrt{2}}|j_1^2; J=0\rangle - \frac{1}{\sqrt{2}}|j_2^2; J=0\rangle. \tag{5.21}$$

The state expressed in equation (5.20) is termed a 'coherent' pair-correlated state and that in equation (5.21) is an 'incoherent' pair-correlated state. These can be compared with the formulation presented in [4], chapter 4, section 4.5.1, as equations (4.85) and (4.86) therein. Equation (5.20) is the essence of the so-called 'Cooper' pair.

A structure similar to that shown in figure 4.21, for ^{146}Gd (and ^{147}Tb), is shown in figure 5.12, for ^{90}Zr (and ^{91}Nb). A test of state vectors for the $|0_1^+\rangle$ and $|0_2^+\rangle$ states in ^{90}Zr can be made with the one-proton stripping reaction, ^{89}Y(^{3}He, d)^{90}Zr and the result is shown in figure 5.13.

The model result, which is manifested in equation (5.20), and the data-based conclusion manifested in figure 5.13 provide an elementary perspective on the structure of Cooper pairs, cf. equations (4.1) and (4.2). As a concept, it leads to the view of the ground state of ^{90}Zr as depicted in figure 5.14, i.e. of a diffuse occupancy of the subshells between $Z = 28$ and $Z = 50$. Indeed, while this diffuseness is concentrated within the major shell filling, it is possible that it extends into neighboring shells. This view is one of the key results of the BCS theory of pairing in nuclei, which allows for many pairs in non-degenerate subshells within a major shell. We defer details of BCS theory, as applied to nuclei, until later in the Series. But figure 5.14 embodies the key result of BCS theory and the essential mathematical structure is presented in [4], chapter 4, section 4.5.3.

From these few details it is evident that the states in ^{89}Y that are labeled as $|2p_{1/2}\rangle$ and $|1g_{9/2}\rangle$ are not simple single-particle states: they exist in an 'environment' of correlated pairs. While the correlated pairs cannot involve the $|2p_{1/2}^2; J=0\rangle$ configuration in the ^{89}Y ground state—this configuration is 'blocked' to pair correlations by the unpaired proton—all of the other shell model configurations in

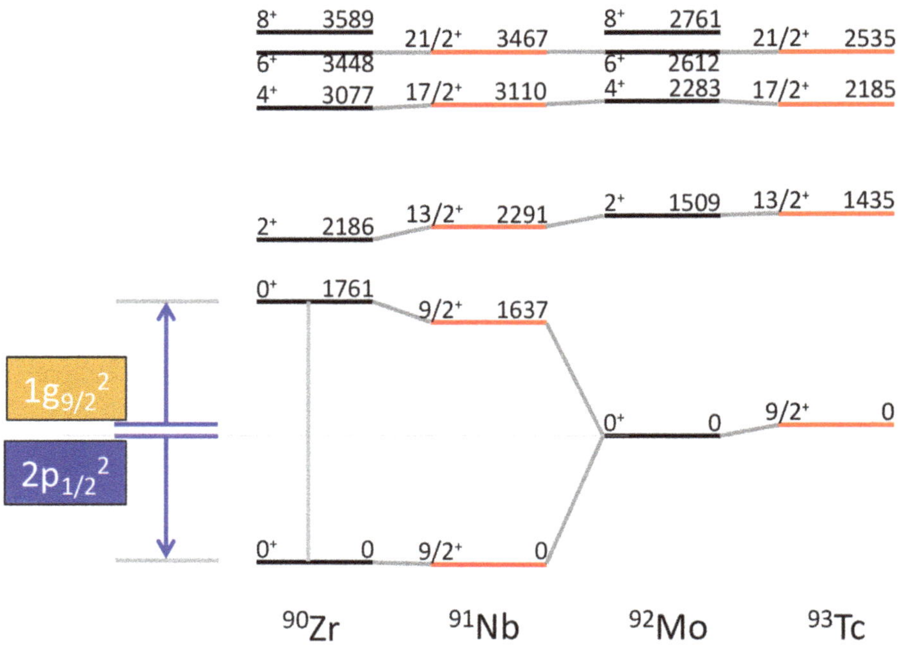

Figure 5.12. Systematics of selected positive-parity states in the $N = 50$ isotones. These states are dominated by the proton configurations $2p_{1/2}$ and $1g_{9/2}$. The appearance of two states with 0^+ in ^{90}Zr (and two states with $9/2^+$ in ^{91}Nb) is explained by mixing of the pair configurations $|2p_{1/2}^2; J = 0\rangle$ and $|1g_{9/2}^2; J = 0\rangle$, as described by the model shown in figure 5.11. This mixture of configurations is supported by data shown in figure 5.13 and 5.15.

the shell play a role in determining the properties of the ^{89}Y ground state. The $|2p_{1/2}\rangle$ configuration is said to be 'dressed' by pairing correlations. A single-particle structure that is dressed in pairing correlations is called a single *quasiparticle*. The blocking effect contributes to the odd–even staggering in one-nucleon separation energies, as appears in figures 4.23 and 4.24.

A direct consequence of the diffuse occupancy of shell model orbits that result from pairing correlations is, e.g. one-proton pickup and stripping reactions on ^{90}Zr populate states in ^{89}Y and ^{91}Nb that would not be observed if the shell model orbitals had a sharply defined occupancy. This is depicted in figure 5.15.

5.3 Rotor models for nuclear structure

Models of nuclear rotation form a very large body of literature in nuclear physics. Herein, the presentation only looks at the simplest possible models, with a focus on their role in the organization of nuclear data.

There are useful analogies between nuclear rotation and molecular rotation. A few simple examples are given.

The most remarkable property of small finite systems is that they may exhibit quantized energies that are relatively independent of their internal many-body structure. This can be associated with collective behavior of the entire system: the behavior can be

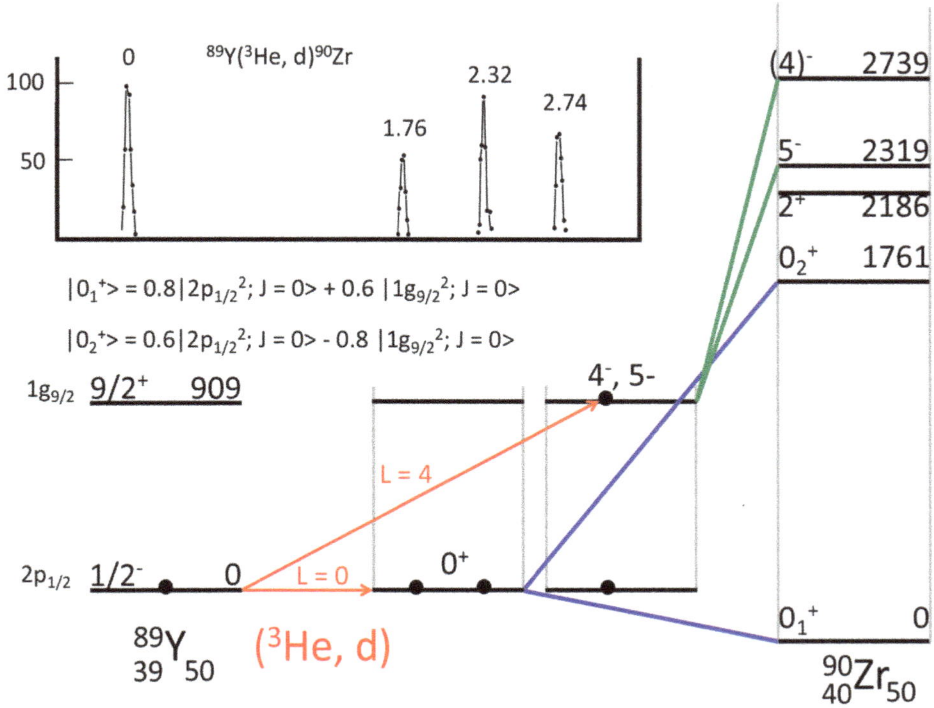

Figure 5.13. The spectrum of deuterons from the proton stripping reaction ^{89}Y(^{3}He, d)^{90}Zr. The spectrum shows significant population of the 0_1^+ and 0_2^+ states in ^{90}Zr. This reveals that the two states must both contain significant amplitudes of the $|2p_{1/2}^2; J = 0\rangle$ configuration, confirming the inference made in figure 5.12 from energy systematics for the $N = 50$ isotones. The inset spectrum is a plot of data presented in [10].

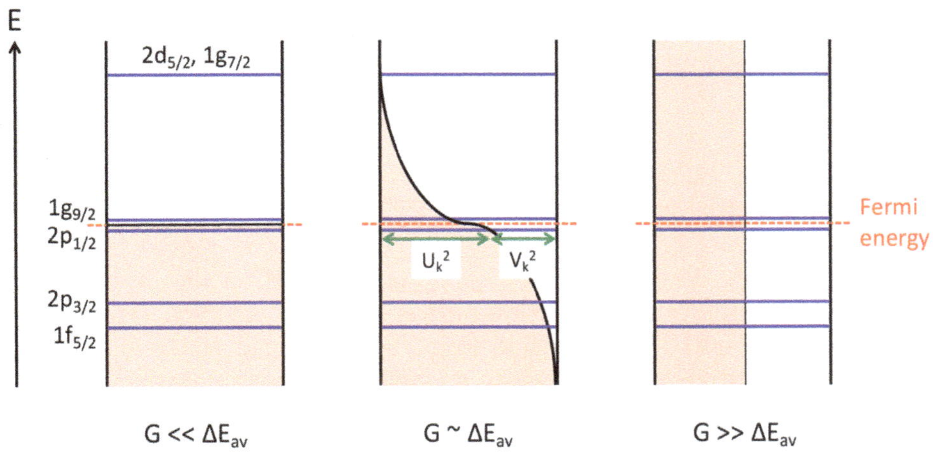

Figure 5.14. A schematic view of orbital occupancy for ^{90}Zr is shown in the middle frame. The left and right frames show the two limits of vanishing pairing correlations and extreme dominance of pairing correlations. The three cases are expressed as pairing force strength, G relative to average spacing between levels. The Fermi energy marks the filling or average filling of the levels. The factors U_k^2 and V_k^2 are measures of the orbital emptiness and occupancy, respectively, and they sum to unity. The variation in orbital occupancy shown is purely schematic; but is typical of pairing strengths in this mass region. The near 50:50 emptiness/occupancy of the ground state of ^{90}Zr is proven by the data shown in figure 5.15.

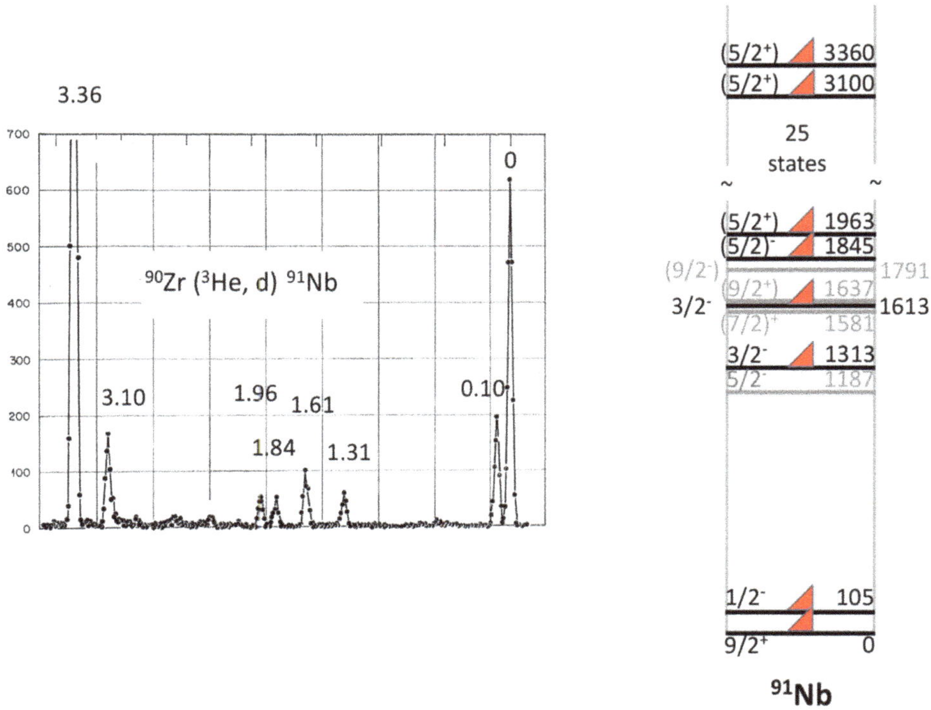

Figure 5.15. The spectrum of deuterons from the proton stripping reaction ^{90}Zr(^{3}He, d)^{91}Nb. The spectrum shows significant population of the $2p_{1/2}$ state in ^{91}Nb, which again contradicts the naïve independent-particle view that this orbital is completely filled in ^{90}Zr and therefore the population of the state by addition of a proton is forbidden by the Pauli principle. Note the selective population of other states in ^{91}Nb, such as the population of $3/2^-$ and $5/2^-$ states, which indicates that the $2p_{3/2}$ and $1f_{5/2}$ orbitals also are not fully occupied. Further note the population of $5/2^+$ states that must involve the $2d_{5/2}$ orbital from above the $Z = 50$ closed shell: the shell energy gap of ~4 MeV, cf. figure 4.12, is reduced by a gain in pairing correlation energy through the 'unblocking' of orbitals near the Fermi energy, cf. figure 5.14. Reprinted with permission from [11]. Copyright (1969) by the American Physical Society.

expressed in terms of the total spin of the body. The resulting excitation spectra provide key information on the 'intrinsic' properties of the body as a whole, e.g. the 'shape' of the body.

In chapter 3, key details of rotor models are presented, in the form of simple equations, which result from the simplest models of nuclear rotation. These treat the nucleus as a rigid droplet of fluid or liquid drop. The term 'liquid' here is to convey that the nucleon substructure of the nucleus is ignored. It does not imply that the droplet will deform when rotating, as would a droplet of water, in response to centrifugal forces. The data presented in chapter 3 suggest that nuclei do not change their deformation when they rotate. But the data clearly show that the concept of a moment of inertia, when applied to nuclei, exhibits features that do not match any classical mechanical limiting case. This is very different from molecules where, e.g. a diatomic molecule such as hydrogen chloride exhibits a moment of inertia that can be accurately estimated from the masses of the two atoms and the distance between them.

5.3.1 The dumbbell rotor model

The dumbbell model of quantum mechanical rotations is the simplest possible model. It is not applicable to nuclei; but it is applicable to all diatomic molecules. It provides some key insights into what is meant by a quantum mechanical rotor model.

The dumbbell model is a two-body problem in three-dimensional space. It can be formulated directly by adapting the quantum mechanics of central force problems, e.g. the hydrogen atom. We start with the Hamiltonian for a two-body central force problem, as presented in [2], chapter 12, viz.

$$H = \frac{p_r^2}{2\mu} + V(r) + \frac{L^2}{2\mu r^2}, \tag{5.22}$$

where μ is the reduced mass of the two-body system,

$$\mu = \frac{m_1 + m_2}{m_1 m_2}, \tag{5.23}$$

and m_1 and m_2 are the masses of the two bodies. Further, p_r is the momentum directed along the line joining the two bodies and describes their relative motion; $V(r)$ describes the interaction potential holding the two bodies together, expressed in terms of the intrinsic degree of freedom, the distance between the two bodies, r; and L describes the angular momentum of the entire system. Figure 5.16 depicts the essential details.

To turn this central-force inspired formulation into the description of a rotating dumbbell, the Hamiltonian can be expressed as

$$H = \frac{p_r^2}{2\mu} + V(r) + \frac{L^2}{2\mu r_0^2} + \left(\frac{L^2}{2\mu}\right)\left(\frac{1}{r^2} - \frac{1}{r_0^2}\right), \tag{5.24}$$

where r_0 is a constant, equal to the ground-state equilibrium distance between the two-bodies. The Hamiltonian for the rotating dumbbell is then given by

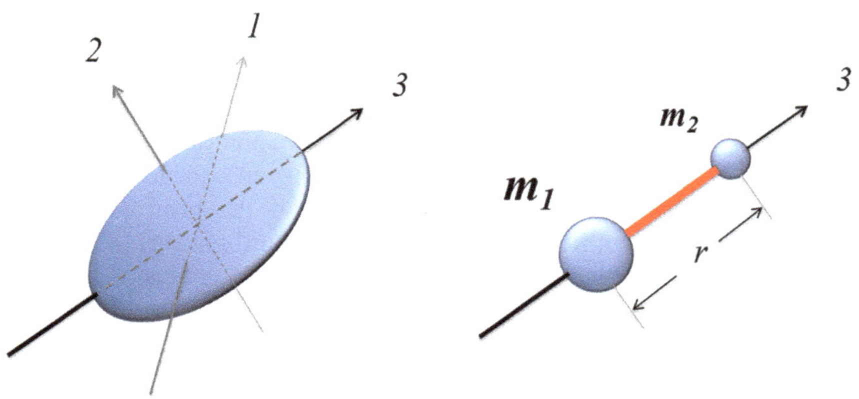

Figure 5.16. Pictorial views of a prolate spheroidal nucleus and a diatomic molecule: both exhibit quantized rotations; but the physics underlying the moments of inertia deduced from rotational energy spectra are very different and are discussed in the text.

$$H_{\text{rot}} := \frac{L^2}{2\mu r_0^2}, \tag{5.25}$$

and the coupling between the rotations and the radial (intrinsic) degree of freedom is given by

$$V_{\text{coupl}} := \left(\frac{L^2}{2\mu}\right)\left(\frac{1}{r^2} - \frac{1}{r_0^2}\right). \tag{5.26}$$

The eigenvalues of H_{rot} can directly be stated, viz.

$$E_{\text{rot}} = AL(L + 1), \tag{5.27}$$

where A is usually termed the rotational energy constant and is seen to conform to the definition of a moment of inertia for the system.

This extraordinarily simple model provides an excellent theoretical framework for interpreting excitation spectra of diatomic molecules. For example, the excitation spectrum of the hydrogen chloride, HCl, molecule is shown in figure 5.17. This is a spectrum of transitions between the rotational band built on the ground state and the rotational band built on the one-phonon excitation, i.e. the first excitation of the intrinsic or inter-atomic degree of freedom. The spectrum shows a 'doubling' of all lines in the spectrum, which reflects the fact that chlorine possesses two stable isotopes with atomic masses of 35 and 37 (with natural abundances of 76% and 24%). The energy 'splitting' of the doublets reflects the differences in the moments of inertia. At a finer level of detail, the rotational energy constants between the ground-state rotational band and that built on the one-phonon excitation are slightly different and are consistent with r_0 being slightly larger for the 'one-phonon' rotational band. Yet further, the spacing of the rotational states deviates from an $L(L + 1)$ pattern as a function of increasing L value and this deviation can be fitted with high precision by adding a term $BL^2(L + 1)^2$ to E_{rot} (see figure 2.12). Such a term, with an estimate of the new parameter B can be derived from a perturbation theory treatment of V_{coupl}.

The dumbbell model forms the basis of looking more deeply into molecular rotations, at the level of intrinsic symmetries. A dramatic example of this is the hydrogen, H_2 molecule. The excited state energy level diagram is shown in figure 5.18. Two bands of rotational states are shown that differ in their parities. This reflects the fact that a proton has a spin of 1/2. In consequence, an H_2 molecule can exist in two distinct configurations: either with the proton spins parallel, so-called 'ortho' hydrogen; or with the proton spins anti-parallel, so-called 'para' hydrogen. The scale at which this originates is of nuclear dimensions, i.e. 10^{-14} m, whereas the inter-atomic distance of the H_2 molecule is $\sim 10^{-10}$ m.

5.3.2 Simple rotor models for nuclei

The simplest rotational models for nuclei can be directly adapted from those used to describe molecules. Many of the essential details are already presented in chapter 3.

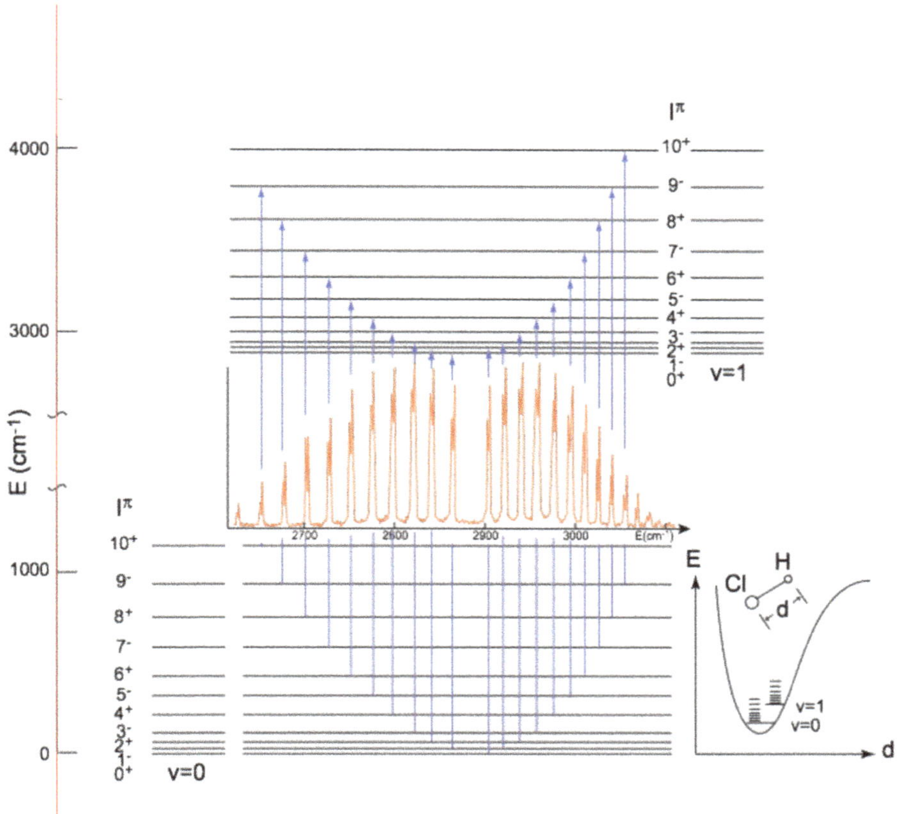

Figure 5.17. Quantized rotations of the hydrogen chloride molecule, HCl as observed by infrared absorption spectroscopy involving the fine structure of the first vibrational excitation of the molecule: this is depicted using the inset in the lower right-hand corner. Note that all absorption lines are shown inverted as if they were emission lines, to fit the perspective of nuclear spectroscopy where study is of emission lines. Note also, all lines are doublets because chlorine possesses two stable isotopes with natural abundances of 76% ^{35}Cl and 24% ^{37}Cl. The energy scale is given in cm^{-1}, 1 cm^{-1} = 1.240 5 $\times 10^{-4}$ eV. The rotational constants for the two bands are 10.437 cm^{-1} (ground-state, $\nu = 0$ band) and 10.134 cm^{-1} (first excited vibrational, $\nu = 1$ band).

Thus, one can discuss moments of inertia of nuclei: albeit, unlike molecules their magnitudes do not match any classical conception of a moment of inertia. Further, one can discuss the intrinsic symmetries possessed by nuclei. Most dramatically, the quantum numbers of the axially symmetric rotor model (cf. figure 3.1) provide an extraordinarily powerful quantum mechanical framework for discussing the electromagnetic properties of nuclei as associated with the nuclear rotational degree of freedom. Herein, we focus on the derivation of important quantum mechanical results for the axially symmetric rotor.

To move beyond the basic $I(I+1)$ pattern of excitation energies and the Clebsch–Gordan coefficients appearing in transition and diagonal matrix elements for electromagnetic properties associated with nuclear rotation, quantum mechanical configuration mixing must be handled. A rotational band is defined by a value of

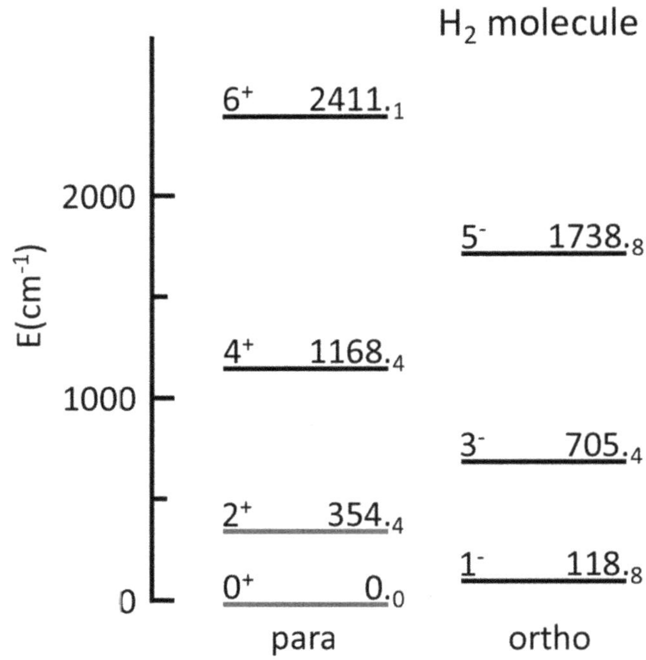

Figure 5.18. The first few energy eigenstates of the rotational spectrum of the hydrogen molecule. Energies are give in cm^{-1}. States are labeled by rotational angular momentum and parity. Parahydrogen involves the two proton spins in a singlet state, $S = 0$, i.e. the spins are anti-parallel; in orthohydrogen they are in a triplet, $S = 1$ state with parallel spins.

the K quantum number and a parity. 'Mixing' may involve configurations with the same or different K values (parity is not mixed). The observed categories of K mixing are $\Delta K = 0$, $\Delta K = 1$, and $\Delta K = 2$, in an obvious notation. Empirically, their origins are from very different physical effects in nuclei. In broad terms, $\Delta K = 0$ mixing may be due to centrifugal stretching or the nuclear pairing interaction; however, experimental data suggest that the origin of this mixing is due to other sources, as yet unidentified: this will be handled later in the Series. The origin of $\Delta K = 1$ mixing appears to be best described as a 'Coriolis' effect. The occurrence of $\Delta K = 2$ mixing appears to be best described as due to axial asymmetry. In the following, we elaborate on this categorization of K mixing effects in nuclei by deriving the mixing matrix elements for $\Delta K = 1$ mixing. Axial asymmetry and $\Delta K = 2$ mixing will be handled later in the Series.

The most widely quantified K mixing in nuclei is $\Delta K = 1$ mixing. It is clearly manifested in the energies of $K = 1/2$ bands observed in odd-mass nuclei. The eigenvectors for $K = 1/2$ bands are expressed (cf. equation (3.5), with M suppressed) as

$$|I, K = 1/2\rangle = \frac{|I, 1/2\rangle + (-1)^{I+1/2}|I, -1/2\rangle}{\sqrt{2}}, \tag{5.28}$$

the term responsible for mixing is (cf. equation (3.14))

$$V_{\Delta K=1} = -\frac{\mathbf{I} \cdot \mathbf{j}}{\mathcal{J}}, \tag{5.29}$$

which leads to the expression (cf. equation (3.16))

$$E = A[I(I+1) + a(-1)^{I+1/2}(I+1/2)], \tag{5.30}$$

where a is a parameter. This relationship follows from the expansion

$$\mathbf{I} \cdot \mathbf{j} = 1/2(I_+ j_- + I_- j_+) + I_z j_z, \tag{5.31}$$

together with the standard action of the 'raising' and 'lowering' operators (cf. [2], chapter 11), viz.

$$I_+|IMK\rangle = \sqrt{(I-K)(I+K+1)}\,\hbar|IMK+1\rangle, \tag{5.32}$$

$$j_+|j\Omega\rangle = \sqrt{(j+\Omega)(j-\Omega+1)}\,\hbar|j\Omega-1\rangle, \tag{5.33}$$

where, note, in the body frame j_+ acts as a lowering operator. For present purposes, we do not give an explanation of this. Similar relationships for I_- and j_-, with the sign change for j_-, are obtained. The eigenvalue of I_z is K and of j_z is Ω. For the low-lying bands in nuclei, $K = \Omega$, and so the interaction produces $\Delta K = 1$ mixing. From these equations, an expression for the parameter a (cf. equation (5.30)) can be obtained. (Note, the $\Delta K = 1$ mixing is between the $K = +1/2$ and $K = -1/2$ amplitudes in the state vector, equation (5.28), i.e. it is within a single $K = 1/2$ band.)

Mixing of bands differing in K by one unit and the impact of such mixing on electromagnetic properties are deferred to later in the Series.

5.3.3 Models of moments of inertia for nuclei

The issue of nuclear moments of inertia remains a puzzle. There is a large literature that attempts to model nuclear moments of inertia. There is not a consensus regarding what is happening when it is said that a nucleus 'rotates'. For example, while 'rotation' is kinetic, it is possible to have interactions that give an energy spectrum that possesses an '$I(I+1)$' dependence. Interactions are potential energy, not kinetic energy. Nevertheless, we like to talk of moments of inertia. A few model recipes for moments of inertia are provided.

The simplest model of a moment of inertia is the classical concept of a rigid body that is rotating. The moment of inertia of a classical body is given by

$$I = \sum_{i=1,\ldots,A} m_i r_i^2. \tag{5.34}$$

For a sphere of uniform density, $I = 2/5MR^2$, where M is the total mass and R is the radius of the spherical body. For an ellipsoidal body, allowance must be made for the departure from a spherical shape and this is done using the parameters β and γ, cf. figure 5.19(a). We note that this parameterization is for a sharp surface of a uniform density droplet. (As presented in chapter 1, nuclei are not of uniform

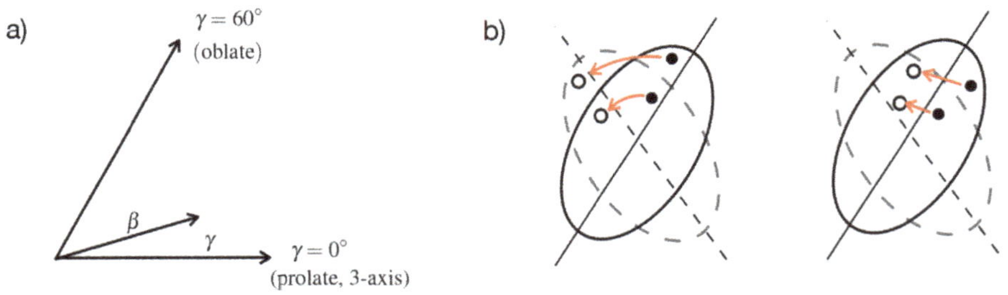

Figure 5.19. (a) Parameters, β, γ used to describe the deformation of an ellipsoidal liquid droplet. (b) Types of rotational flow in an ellipsoidal body. Details are discussed in the text.

density—they possess a surface diffuseness, cf. figure 1.3—but the present formulation is sufficient for discussion.)

A model of moments of inertia that is more subtle involves what is termed 'irrotational flow'. Imagine an ocean planet observed from space, with bi-lobal unbroken water waves circulating the planet. We know from experience that an unbroken water wave only involves motion transverse to the water surface, there is no lateral motion. Therefore, from a distance, the ocean planet appears to be rotating, but nothing is moving laterally. This view can be converted to the concept of an oval shaped bucket containing a superfluid and rotating. The fluid adjusts to the constraint of its container (the bucket), but it does so by a more radial-like flow. (Strictly speaking, a superfluid cannot support viscosity and so it can only respond by moving perpendicularly to the surface of the bucket.)

Figure 5.19(b) provides sketches of rigid-body and irrotational flow. The derivation of moments of inertia for an ellipsoidal body of uniform density are obtained as follows. For a nucleus with

$$R(\theta, \phi) = R_0[1 + g + \beta Y_{20}(\theta, \phi)],\ g \approx -\frac{\beta^2}{4\pi}\ \text{(volume conservation)} \tag{5.35}$$

$$Q_0 = \frac{3}{\sqrt{5\pi}} ZeR_0^2\beta\left(1 + \frac{1}{8}\sqrt{\frac{5}{\pi}}\beta + \cdots\right). \tag{5.36}$$

For an axially asymmetric quadrupole shape:

$$R(\theta, \phi) = R_0\left[1 + \beta\left\{\cos\gamma Y_{20} + \frac{\sin\gamma}{\sqrt{2}}(Y_{22} + Y_{2-2})\right\}\right], \tag{5.37}$$

$$\delta R_1 = R\left(\frac{\pi}{2}, 0\right) - R_0 = R_0\beta\sqrt{\frac{5}{4\pi}}\cos\left(\gamma - \frac{2\pi}{3}\right), \tag{5.38}$$

$$\delta R_2 = R\left(\frac{\pi}{2}, \frac{\pi}{2}\right) - R_0 = R_0\beta\sqrt{\frac{5}{4\pi}}\cos\left(\gamma + \frac{2\pi}{3}\right), \tag{5.39}$$

$$\delta R_3 = R(0, 0) - R_0 = R_0\beta\sqrt{\frac{5}{4\pi}}\cos\gamma. \tag{5.40}$$

For rigid flow:

$$\mathcal{J}_1^{\text{rig.}} = \frac{2}{5}AMR_0^2\left\{1 - \sqrt{\frac{5}{4\pi}}\beta\cos\left(\gamma - \frac{2\pi}{3}\right) + \cdots\right\}, \tag{5.41}$$

$$\mathcal{J}_2^{\text{rig.}} = \frac{2}{5}AMR_0^2\left\{1 - \sqrt{\frac{5}{4\pi}}\beta\cos\left(\gamma + \frac{2\pi}{3}\right) + \cdots\right\}, \tag{5.42}$$

$$\mathcal{J}_3^{\text{rig.}} = \frac{2}{5}AMR_0^2\left\{1 - \sqrt{\frac{5}{4\pi}}\beta\cos\gamma + \cdots\right\}. \tag{5.43}$$

For irrotational flow:

$$\mathcal{J}_1^{\text{irrot.}} = \frac{3}{2\pi}AMR_0^2\left\{\beta^2\sin^2\left(\gamma - \frac{2\pi}{3}\right) + \cdots\right\}, \tag{5.44}$$

$$\mathcal{J}_2^{\text{irrot.}} = \frac{3}{2\pi}AMR_0^2\left\{\beta^2\sin^2\left(\gamma + \frac{2\pi}{3}\right) + \cdots\right\}, \tag{5.45}$$

$$\mathcal{J}_3^{\text{irrot.}} = \frac{3}{2\pi}AMR_0^2\{\beta^2\sin^2\gamma + \cdots\}. \tag{5.46}$$

The global situation for moments of inertia for nuclei is presented in figure 3.27. All nuclei exhibit moments of inertia between rigid-body and irrotational flow. This has attracted extensive model building based on a rotating superfluid, with suitable blocking effects of the pairing correlations underlying the superfluid to arrive at a less-than superfluid flow. However, the behavior of rotational energies in even-mass nuclei and their odd-mass neighbors, as presented in figure 3.9(a)–(d), contradicts such a view in a major way, i.e. the rotational profiles are identical and there is not a blocking effect due to the unpaired nucleon. At present it must be concluded that we do not know how nuclei rotate.

5.4 The Bohr model for nuclear structure

Second only to the shell model, the second longest-standing model of nuclear structure is the Bohr model. This is a liquid-drop model with quantized rotations and vibrations. It has passed through a number of developmental phases. In the hands of Bohr and Mottelson, its adiabatic limits were explored; this is often referred to as the Bohr–Mottelson model. An exploration in greater depth was made by the Frankfurt Group, and is referred to as the Bohr–Mottelson–Frankfurt model. The full diagonalization was achieved by the Toronto Group; and is called the algebraic collective model (ACM). Herein, we limit details to just a basic introduction to the adiabatic limits of the basic Bohr model.

The Bohr model is a quantum mechanical description of a liquid droplet. It can be viewed from two perspectives: that of the qualitative behavior to be expected—multipole modes exhibiting vibrations and rotations; and that of the quantum mechanical treatment of the leading-order, quadrupole mode. The first perspective is pictorial. The second requires quantum mechanics at the intermediate–advanced level, and an introduction is given in [4], chapter 6.

The adiabatic limit can be 'imposed' by freezing the liquid drop so that it has a rigid deformation (including the possibility of being spherical). The frozen object will possess a set of intrinsic symmetries, e.g. axial symmetry and reflection symmetry. This 'freezing' can be relaxed by allowing vibrations about equilibrium 'shapes', these will be so-called normal-mode vibrations. A critical consideration in all of this is the adoption of a body-frame view of the nucleus. This means that the orientation degree of freedom has been factored out of the description. This leads to a simplification of the depiction and to a simplification of the mathematical description of the nucleus. It is also a step to a major caution where observations are concerned: measurements are made in the laboratory frame of reference and do not provide a 'sharp' view of the intrinsic details of a nucleus. This is formalized in terms of rotationally invariant quantities which 'look the same' in the body frame and in the laboratory frame. The simplest rotationally invariant quantity is a length —$\mathbf{r} \cdot \mathbf{r}$ looks the same viewed in the body frame as it does viewed as a rotating object from the laboratory frame. This key concept of deformed objects, for quadrupole deformation, is formalized in the so-called Kumar–Cline sum rules, which is a topic deferred until advanced aspects of nuclear structure are handled.

For present purposes, we confine the discussion to the adiabatic limit of the Bohr model, i.e. vibrations about a near-rigid deformed shape. We limit this to deformed shapes that are axially symmetric. There are two possible normal modes, one gives rise to $K = 2$ bands and the other to $K = 0$ bands; they are referred to as gamma and beta vibrations, respectively. They should possess enhanced $B(E2)$ electromagnetic decay strength to the ground state $K = 0$ band. Essential details are depicted in figure 5.21.

Bands with $K = 2$ and enhanced $E2$ decay to the ground-state band are observed. The band-head state with spin-parity 2^+ is observed to decay to the ground state with a $B(E2)$ that is typically 5 W.u. An example is shown in figure 3.18. Bands with $K = 0$ and enhanced $E2$ decay to the ground-state band are not observed as a systematic feature of deformed nuclei. One would naturally expect that, since the two modes are normal modes of collectivity for the nucleus behaving as a quantized liquid drop, if one mode exists then so should the other. This casts doubt on the interpretation of the $K = 2$ bands as vibrational modes of a liquid drop. Further discussion is beyond the present scope of detail.

The existence of $K = 2$ bands with collective $E2$ decay to the ground-state band can also occur for nuclei that are ellipsoidal: these can be described using models of a triaxial rotor. Indeed, there are nuclei that exhibit $K = 2$ bands that possess moderately strong $E2$ decays of this type, but they are not so strongly deformed overall. A triaxial rotor possesses excited $K = 4$ bands that have collective $E2$ decay to the $K = 2$ bands. Peculiarly, for many of the nuclei that are candidates for exhibiting triaxial rotor behavior, the candidate $K = 4$ bands possess hexadecapole character, i.e. they have direct $E4$ decay to the ground state. Yet further, these $K = 4$ bands are populated in one-nucleon transfer reactions. The present status of these

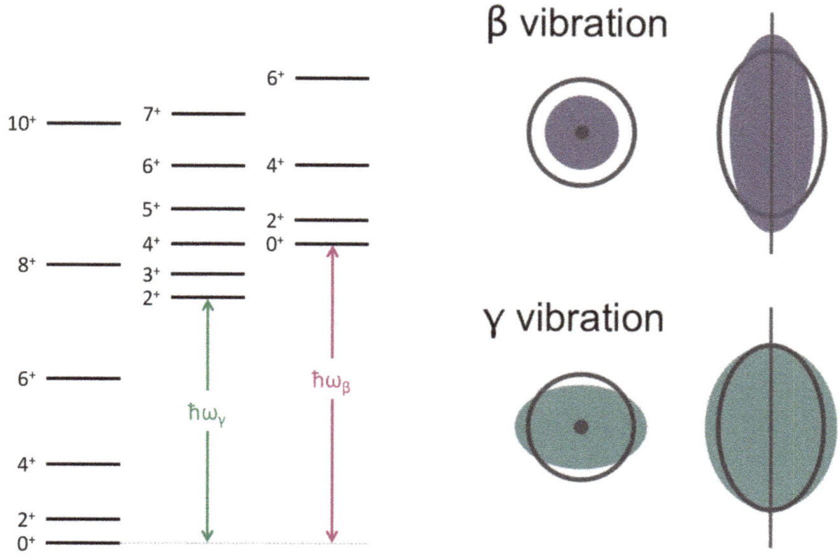

Figure 5.20. The normal modes of shape vibration in a spheroidal body and the bands associated with one-phonon excitations.

issues is most unsatisfactory, and has received very little attention in the past 30 years. It will be handled in detail later in the Series.

As a final topic in this direction for deformed nuclei, there is clear evidence for weak octupole collectivity in such nuclei. This may extend to octupole deformation. There are observations that are consistent with octupole deformation; but direct proof would require the observation of octupole moments and such moments cannot be observed. The reason for this non-observable aspect of octupole deformation is because the octupole moment operator has odd-parity: as with dipole operators (which have odd parity), diagonal matrix elements are zero. Thus, all inferences about octupole deformation are model based. As with candidate nuclei for triaxial deformation, so candidate nuclei for octupole deformation are not so strongly deformed overall.

For reference purposes, schematic views of vibrations and the associated band structures expected from the adiabatic Bohr model are shown in figures 5.20 and 5.21. Figure 5.22 shows the band structures expected for a triaxial rotor.

There remains the issue of vibrations in spherical nuclei. This degree of freedom has been the subject of a complete upset in views in the past fifteen years. It appears that such degrees of freedom do not exist. The failure is in the absence of candidate states for multi-phonon excitations. While energy patterns had convinced the nuclear physics community that this degree of freedom was present in nuclei, $B(E2)$ patterns (which are much harder to elucidate) do not support quadrupole vibrations in spherical nuclei. Indeed, spherical nuclei appear to be very rare. For reference purposes, the excitation pattern with associated $B(E2)$ values for quadrupole vibrations in a spherical nucleus is shown in figure 5.23.

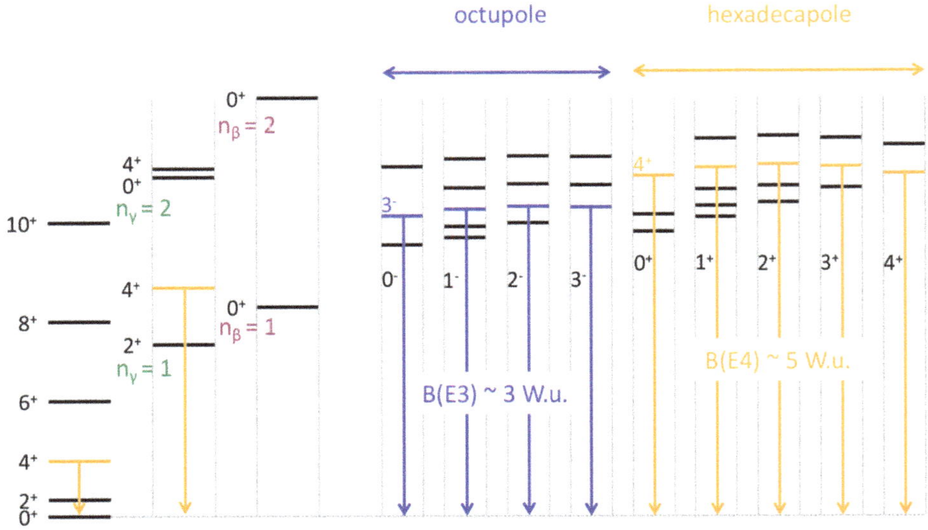

Figure 5.21. A multi-band view of excitations expected for a liquid-drop model of the nucleus. This includes multi-phonon beta and gamma vibrations, octupole vibrations and hexadecapole vibrations.

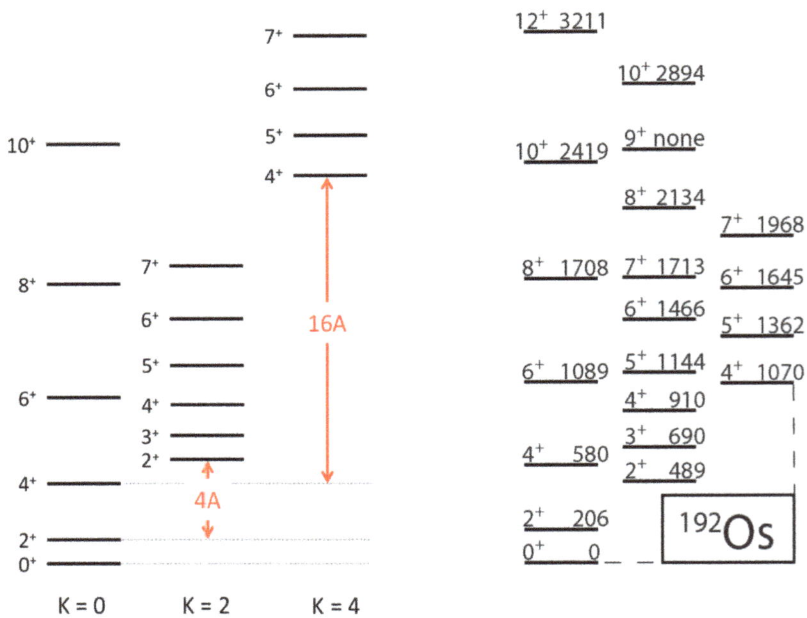

Figure 5.22. Band structures expected for a triaxial rotor and (right-hand side) bands observed in ^{192}Os. The energy spectrum for a triaxial rotor can be approximated by the relationship $E(I, K) = BI(I + 1) + AK^2$.

The message that emerges, from the above brief summary, is that the occurrence of shape coexistence and collectivity associated with weak deformation, especially triaxial deformation, need closer attention and they are on-going primary research concerns. As such, we defer a deeper look into these issues until forthcoming books in the Series.

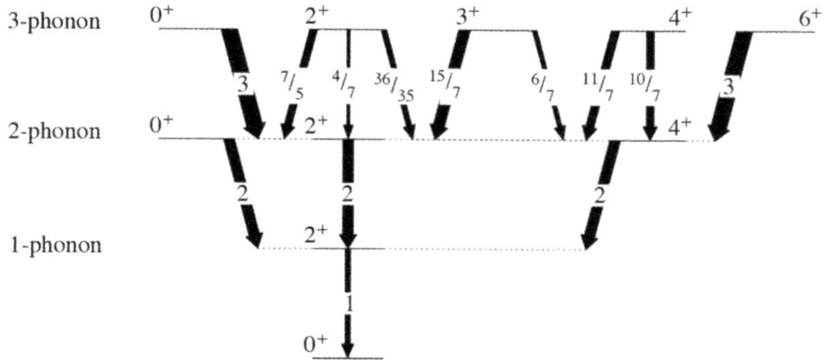

Figure 5.23. Spectrum of a harmonic quadrupole vibrator. The energies of excited states are multiples of the 1-phonon state energy. The electromagnetic transition strengths for n-phonon → $(n-1)$-phonon states are multiples of the 1-phonon → 0-phonon transition strength. The 3-phonon transition strengths, where fragmented to multiple final states in the 2-phonon multiplet, sum to three units. The details of the branching ratios are beyond the present level of discussion. We note that in this model, all quadrupole moments are zero. The allowed spin values for the multi-phonon states can be determined using the m scheme for $L = 2$ bosons, cf. section 4.2.

5.5 Exercises

The relative challenge and time investment associated with the exercises is indicated from * to ***. Exercises marked with '**E**' require interactive access to figures or watching included video exercises and will only have such functionality in the e-book version.

5-1. With reference to figure 5.10, make a similar plot for $\epsilon_4 < 0$. *

5-2. Considering equations (5.16)–(5.18) and figure 5.11, write down and solve the corresponding equations for the energy eigenvalues for:
   (a) one pair in a three-level system; **
   (b) one pair in a four-level system. **

5-3.
   (a) Calculate the moment of inertia for the hydrogen atom. **
   (b) Calculate the energy in electronvolts of an $I = 1$ rotational state for this system. (Treat the H atom as a classical dumbbell with separation of the proton and the electron given by the Bohr radius of the 1s orbit. Take the radius of the proton as 0.84 fm.) **

5-4. Illustrate the difference in rotational energies for nuclei of mass A and $A + 1$, as depicted in plots of E_γ versus I_i. Use the classical expression for the moment of inertia of a sphere. **

5-5. For the spin sequence, $I = 0, 1, 2, 3, \ldots$, what will E_I/E_1 be for $I = 2, 3, 4$? (cf. equation (2.2)). *

5-6. For the HCl molecules with ^{1}H and ^{35}Cl, ^{37}Cl, the information in figure 5.24 and table 5.1 is available from infrared absorption spectroscopy.
   (a) What is $\hbar\omega$ (cm^{-1}) for the $\nu = 1$ vibrational mode in H^{35}Cl and H^{37}Cl? *

(b) What is the moment of inertia for the $\nu = 0$ ground state in $H^{35}Cl$ and $H^{37}Cl$? *
(c) What is the moment of inertia for the $\nu = 1$ vibrational state in $H^{35}Cl$ and $H^{37}Cl$? *
(d) What is the interatomic distance between the H atom and the Cl atom in these molecules? **
(e) With reference to figure 2.12, check the location of the data point shown for the HCl molecule. *

5-7. For the HCl molecule, the intermolecular potential can be described by the model expression

$$V(R) = D_e[\exp{-2\alpha(R - R_0)} - 2\exp\{-\alpha(R - R_0)\}], \qquad (5.47)$$

where $D_e = 4.62$ eV, $\alpha = 1.874$ nm^{-1}, and $R_0 = 0.127$ nm. It is an example of a so-called 'Morse' potential. Make a drawing to scale of this potential—Wolfram Alpha could be used for this. **

5-8. Using the spectrum for the H_2 molecule shown in figure 5.18, sketch similar energy diagrams for the molecules:
(a) H–D; *
(b) H–T; *
(c) D–D; *

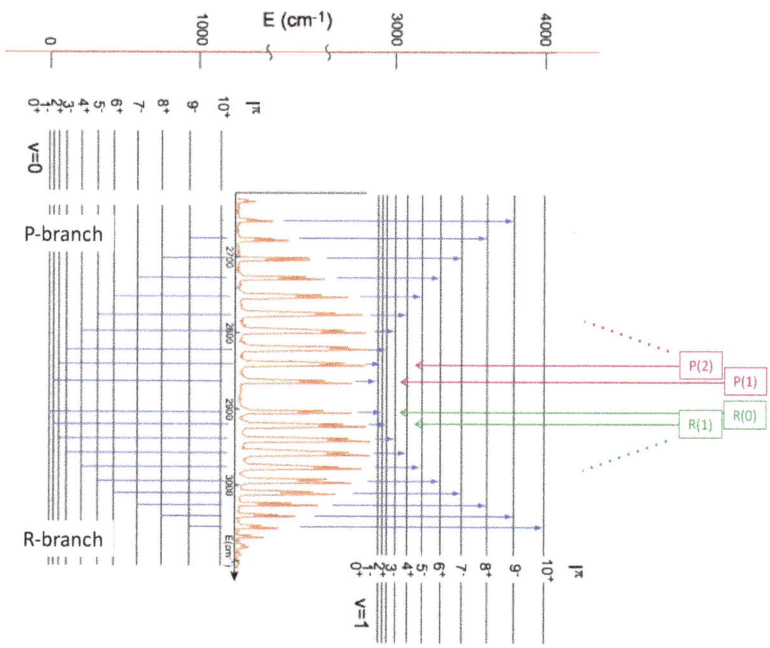

Figure 5.24. A labeled version of figure 5.17 showing the so-called R and P series of infrared absorption lines, cf. table 5.1. Each transition in the R series is labeled by the spin J of the state from which the absorption occurs, the final state has spin $J = J + 1$. The P series absorption involves $J \rightarrow J - 1$ transitions, and so starts with $J = 1$.

Table 5.1. Energies in cm^{-1} for absorption lines in the infrared spectrum of HCl, for the ^{35}Cl and ^{37}Cl isotopes. The column labeling, viz. R(*J*) and P(*J*), where *J* is the spin of the initial state before absorption, is defined in figure 5.24.

	$H^{37}Cl$		$H^{35}Cl$	
J	R(*J*)	P(*J*)	R(*J*)	P(*J*)
0	2 904.08		2 906.25	
1	2 923.71	2 862.99	2 925.9	2 865.09
2	2 942.7	2 841.56	2 944.92	2 843.62
3	2 961.05	2 819.54	2 963.29	2 821.55
4	2 978.74	2 796.95	2 981.01	2 798.92
5	2 995.76	2 773.81	2 998.05	2 775.74
6	3 012.11	2 750.12	3 014.42	2 752.01
7	3 027.76	2 725.92	3 030.1	2 727.76
8	3 042.72	2 701.21	3 045.07	2 702.98
9	3 056.92	2676	3 059.32	2 677.71
10	3 070.51	2 650.28	3 072.85	2 651.94
11		2 624.13	3 085.66	2 625.72
12				2 598.99

5-9. The symmetric rotor with a plane of reflection symmetry and a ground-state spin of zero only has even-spin rotational states. Therefore, it is possible to conduct a search for exotic stable particles in Nature that lead to rare species of molecule that exhibit odd-spin rotational states. What has been done to explore this idea? ***

References

[1] Halcrow C and Harland D 2020 Attractive spin-orbit potential from the Skyrme model *Phys. Rev. Lett.* **125** 042501

[2] Heyde K and Wood J L 2019 *Quantum Mechanics for Nuclear Structure, Volume 1: A primer* (Bristol: IOP Publishing)

[3] Rowe D J and Wood J L 2010 *Fundamentals of Nuclear Models: Foundational Models* (Singapore: World Scientific)

[4] Heyde K and Wood J L 2020 *Quantum Mechanics for Nuclear Structure, Volume 2: An intermediate level view* (Bristol: IOP Publishing)

[5] Schiffer J P *et al.* 2013 Valence nucleon populations in the Ni isotopes *Phys. Rev.* C **87** 034306

[6] Reinhard P G and Otten E W 1984 Transition to deformed shapes as a nuclear Jahn–Teller effect *Nucl. Phys.* A **420** 173

[7] Firestone R B, Baglin C M and Frank Chu S Y 1999 *Table of Isotopes* (New York: Wiley)

[8] Quentin P and Babinet R 1970 Nilsson's Hamiltonian in the asymptotic basis *Nucl. Phys.* A **156** 365

[9] Kleinheinz P, Casten R F and Nilsson B 1973 Hexadecapole deformation in the Wolfram nuclei *Nucl. Phys.* A **203** 539

[10] Vourvopoulos G and Fox J D 1969 Experimental study of the $T = T_z$ states in ^{90}Zr via a (^{3}He,d) reaction *Phys. Rev.* **177** 1558
[11] Cates M R, Ball J B and Newman E 1969 Nuclear-structure studies of ^{91,92,93,95,97}Nb and ^{94}Mo with the (^{3}He,d) reaction *Phys. Rev.* **187** 1682
[12] Gustafsson C, Lamm I-L, Nilsson B and Nilsson S G 1966 Nuclear deformabilities in the rare-earth and actinide regions with excursions off the stability line and into the super-heavy region *Ark. Fys.* **36** 613

Chapter 6

Nuclear spectroscopic measurements

Selected data sets are used to illustrate the variety of spectroscopic measurements that are accessible. The nature of quantum measurement, as manifested using particle and radiation detectors, is commented on. Measurement types are presented in a manner that helps identify what information is obtained, especially limitations. Special attention is given to gamma-ray spectroscopy because it is the dominant means for detection of events. This extends to essential details of the efficiency of standard gamma-ray detectors. Some details of semiconductor detectors for electrons are outlined. Types of reaction spectroscopy involving one and few nucleon transfer are illustrated. Some aspects of atomic spectroscopy pertinent to the study of nuclei are given. Illustration is given of some exotic atom studies related to nuclei. The role of nuclear mass measurements for nuclear structure is sketched. The special topic of on-line study of rare isotopes is given some attention.

Concepts: quantum measurement, safe (Coulomb excitation) energy, Doppler correction, decay standards, pandemonium effect, summing, Compton suppression, Poisson statistics, feeding, draining, line shape, escape peaks, entrance channel, exit channel.

The spectroscopy of nuclei possesses a unique set of techniques, in addition to techniques that have parallels in other branches of physics under the title of 'spectroscopy'. Herein we give some details of how the measurements are conducted, especially with a connection to the data presented in chapters 1–4.

6.1 A quantum measurement preamble

At the heart of nuclear spectroscopy, one is making a *measurement on a quantum mechanical system.* There is a step in the measurement process that has no classical analog, namely that the outcome of the measurement process is in general inherently uncertain. There is a 'spectrum' of outcomes, each with a well-defined probability of observation (excepting instrumental uncertainty).

doi:10.1088/978-0-7503-2674-2ch6

Formally, one expresses quantum mechanical processes as time-dependent changes in stationary states (energy eigenstates). The study of energy eigenstates dominates quantum mechanics, its teaching and its practice. Time-dependent quantum mechanics is rarely developed in an introductory level course. The key to the conceptualization of time-dependent quantum mechanics is to introduce an interaction term into the Hamiltonian. The essence of what this is doing is that it *changes* the Hamiltonian; in consequence, the Hamiltonian is no longer the one that controlled the stationary states of the system, and whatever stationary state the system occupied initially is no longer an energy eigenstate. Thus, in general one observes one of the energy eigenstates of the new Hamiltonian, with a certain probability. Repeating this measurement many times (experimental 'statistical' information or data), one builds up a probability catalog of all possible outcomes. With sufficient experimental data (statistics), one observes even very rare possible outcomes: 'weak decay branches'.

Population of an excited stationary state can be 'active', e.g. Coulomb excitation or inelastic scattering using the interaction of a beam of particles on a target; or it can be 'passive' as in radioactive decay (of a species produced in some prior process). The interaction Hamiltonian will have a time dependence that is manifest if a beam is being used: the interaction will have a range and as the beam particle passes by the target nucleus the interaction strength will increase and then decrease. By integrating over time (distance), the effect of the interaction as a time-dependent Hamiltonian can be quantified.

Conceptualizing the interaction Hamiltonian for radioactive decay is a more subtle issue. For example, decay by γ-ray emission involves the coupling of the nucleus to the electromagnetic (EM) field, the EM 'vacuum': the EM vacuum is always present! This is handled conceptually by separating the observational process into a prior excitation process, e.g. population of the energy eigenstate by interaction with a beam; and a de-excitation process where from the instant that the state is populated, the EM field is suddenly turned on at a constant strength, the interaction strength. Beta decay is conceptualized in a similar manner with respect to the weak field vacuum.

Whether or not a quantum transition process occurs is dictated by conservation of energy. For example in alpha decay, the full 'spectrum' of energy eigenstates includes the ground-state of the parent nucleus and a range of energy eigenstates of the daughter nucleus plus a discrete set of kinetic energies for the alpha particle (plus recoil energies of the daughter nucleus): with the provision that the atomic mass of the parent nuclear species must be greater than the sum of the atomic masses of the daughter nuclear species and a helium atom. The decay process is 'hindered' by the Coulomb barrier through which the alpha particle must 'tunnel', together with 'overlap' of the ground energy eigenstate of the parent nucleus and the various energy eigenstates populated in the daughter nucleus. Thus, one arrives at a probability for the occurrence of alpha decay of a given nucleus. Observation of many decays builds up a probability catalog. Observing the time delay for occurrence of a de-excitation event for a given ensemble of parent nuclei, the characteristic half-life of the decaying state is determined. Decay branches lead to the partial half-lives for the individual decay branches and hence the decay probability.

An issue with respect to the measurement process, which is essentially never discussed, is the actual detection of the radiation. At the heart of the detection process is a quantum mechanical (time dependent) process. One can think of this in terms of the radiation coming into the proximity of the detector and so the situation with respect to the detector Hamiltonian changes and there is a certain probability for the detector energy eigenstate to change. Where detectors are concerned, a chain of amplification is incurred or a 'permanent' change in some medium, e.g. photographic plate exposure, is incurred. These are irreversible processes and the chain of quantum uncertainty comes to an end. It has become popular to speak of quantum 'jumps' in the measurement process—there may be (near) sudden changes in the Hamiltonian controlling the system, followed by the various uncertain outcomes for the system being in an energy eigenstate of the new Hamiltonian—measurement is just an accumulation of entries into a probability catalog. A spectrum is such a catalog, 'binned' with respect to energy, the counts, e.g. per data acquisition channel, reflecting the cumulative probabilities of the nuclear process or sequence of processes leading to that *event*.

The types of measurement that constitute spectroscopy for nuclear structure study can be roughly grouped into categories. These categories have some clearly defined characteristics of procedure. Independently of these characteristics, there are some widely used components to spectroscopic systems such as use of semiconductor detectors. We address some leading aspects of spectroscopy encountered in nuclear structure study.

6.2 Coulomb excitation

A basic and simple (quantum mechanical) way to study nuclei is by Coulomb excitation. This was recognized soon after the quantized nature of nuclear excitation had been proposed in the form of the Bohr collective model [1]. It is basic in the sense that the initial state is well defined—it is the ground states of the target and projectile nuclei. It is a process that is simple—it is the electromagnetic interaction. This can be treated as 'weak' and therefore time-dependent perturbation theory can be used. The information obtained is of fundamental interest—the nuclear response is through the largest electromagnetic matrix elements connected (directly and, in multi-step excitation, indirectly) to the ground states. Observation is via detection of γ-rays emitted as the nucleus 'relaxes' back to its ground state (or an isomeric state) by spontaneous emission from the populated excited states. The population of excited states, determined via γ-ray intensities and location via coincidence relations between pairs, triples, etc., labeled by their energies, leads to an excitation scheme. This can be de-convoluted into a set of electromagnetic matrix elements for transitions between the states observed.

In more recent times it has become possible to obtain precision information by Coulomb excitation of both the target nucleus and the projectile nucleus. This is possible because corrections to γ-ray energies resulting from Doppler shifts can be applied in the 'on-line' acquisition mode. This is done by finely granulated charged-particle detection 'arrays' that can be used to accurately reconstruct the kinematics of the scattered projectile nucleus and the recoiling target nucleus. Indeed, because isotopic purity of a target nuclear species is never 100% and because accelerators function as superb mass separators, it is often the projectile that is the nuclear species under study. This ensures minimization of 'background' γ-ray lines in the spectrum.

The number of excitation steps can be controlled by choice of Z for the nucleus being employed to provide the Coulomb field that is the excitation 'agent' in the experiment. Generally, a low-Z choice leads to one-step and few-step excitations; a high-Z choice leads to multi-step excitations. This can be understood in part in terms of the strength of interaction as a function of Ze and in part, in terms of the duration of close proximity between the target and projectile nuclei: for a given energy, a heavy ion projectile, in the center-of-mass frame, the interaction is for a longer time.

Ideally, large arrays of γ-ray detectors are used (up to 100+ detectors). Not only does this maximize detection efficiency, it also provides 'granularity' for determining angular distributions of detected γ-rays from which spin assignments can be made to excited states. An example is presented in figure 6.1(a) and (b). This example is taken from a data set for Coulomb excitation of states in ^{152}Sm studied as a beam incident on a target of ^{208}Pb. The ^{208}Pb target was highly enriched and, with a first excited state at 2.6 MeV, contributed only weakly to contaminating γ-rays in the spectra.

One-step excitation is used to determine the largest matrix elements involving the ground state of the nucleus under study. For even–even nuclei, this will be the 2_1^+ state, and more weakly the 3_1^- state. This occurs through the $E2$ and $E3$ electromagnetic magnetic matrix elements, which are the observed dominant matrix elements involved in Coulomb excitation of nuclei. (Magnetic matrix elements, e.g. $M1$, are suppressed in Coulomb excitation. This is a topic beyond the present level of treatment of nuclear spectroscopy.) An example of one-step $E2$ excitation is presented in figure 6.2. This example is taken from data sets for Coulomb excitation of 2_1^+ states in the stable even-mass Sn isotopes studied as beams incident on targets of natural carbon and natural titanium. Natural carbon is 98.89% mass 12 and 1.11% mass 13, first excited states are at 4.4 MeV and 3.1 MeV, respectively; natural titanium has masses 46, 47, 48, 49, 50, occurring with 8.25%, 7.44%, 73.72%, 5.41% and 5.18% abundances, respectively, with first excited states, $E(I^\pi)$—889(2^+), 159($7/2^-$) [gs, $5/2^-$], 984(2^+), 1383($3/2^-$) [gs, $7/2^-$], and 1554(2^+), energies in keV, respectively.

A key aspect of Coulomb excitation study is that the center-of-mass energy of the target-projectile MUST remain below the threshold for strong interactions. Because the strong interaction is short-ranged, this is easy to ensure. Always, a 'safe' margin is employed and the energy chosen is referred to as having been conducted at 'safe energies'. The distance of closest approach is determined by the Coulomb repulsion potential energy equated to the selected beam energy; it also has a dependence on the scattering angle: these issues lie beyond the present level of detail. A useful rule is that the distance of closest approach must be at least 5 fm.

The beam energy on the surface of the target facing the beam is the safe energy. As the beam particles pass through the target, they will lose energy through interactions with atomic electrons and then nuclei in the atoms of the target material. It is essential to account for the decreasing energy of beam particles in the target in order to calculate the Coulomb interaction probability.

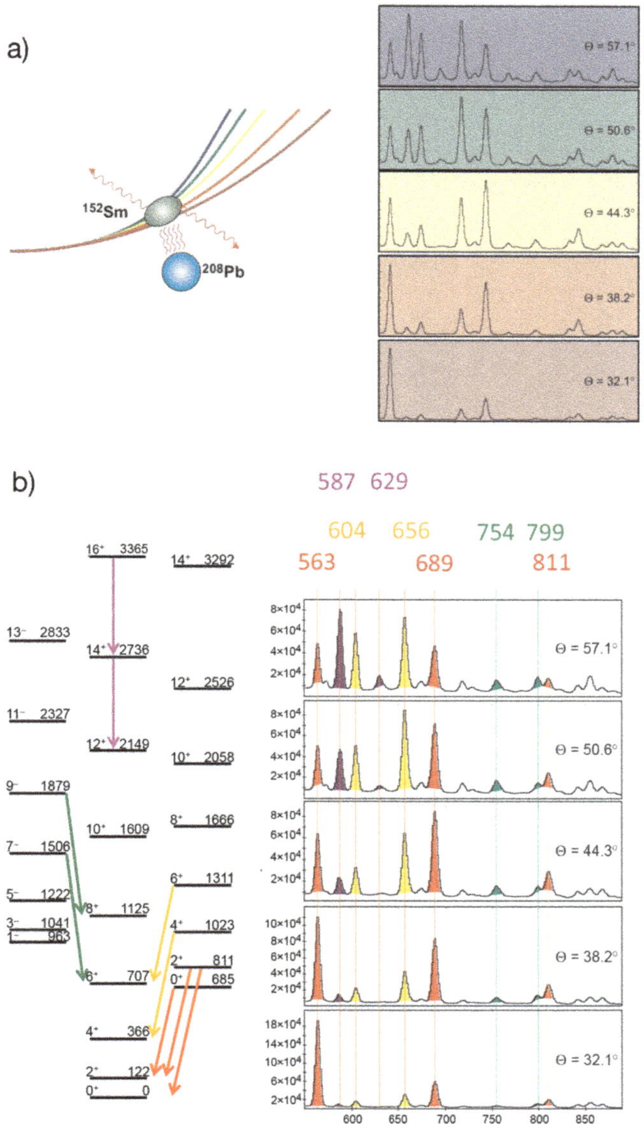

Figure 6.1. Coulomb excitation of the nucleus ^{152}Sm. This shows the variation of γ-ray line intensity as a function of the scattering angle of the projectile with respect to the incident beam direction. There is a high dependence on scattering angle for population of excited states as a function of spin. In (a) the scattering angle of the ^{152}Sm projectile is related to five γ-ray spectra. In (b) the γ-ray lines in these spectra are identified by the corresponding transitions in the ^{152}Sm level scheme (and cf. figure 2.1). The color coding of the γ-ray lines identifies low-spin (red), medium-spin (yellow and green), and high-spin (violet) population. The excitation processes involve multiple steps of *E*2 and *E*3 electromagnetic interaction via the Coulomb field between the ^{152}Sm projectile and the ^{208}Pb target. The de-excitation processes are by spontaneous radiative decay via *E*1 and *E*2 electromagnetic multipoles. The *E*1 and *E*3 multipoles are parity changing. The peculiar roles of *E*3 excitation and *E*1 de-excitation are discussed in the text. The figure is adapted from one that was supplied courtesy of W D Kulp.

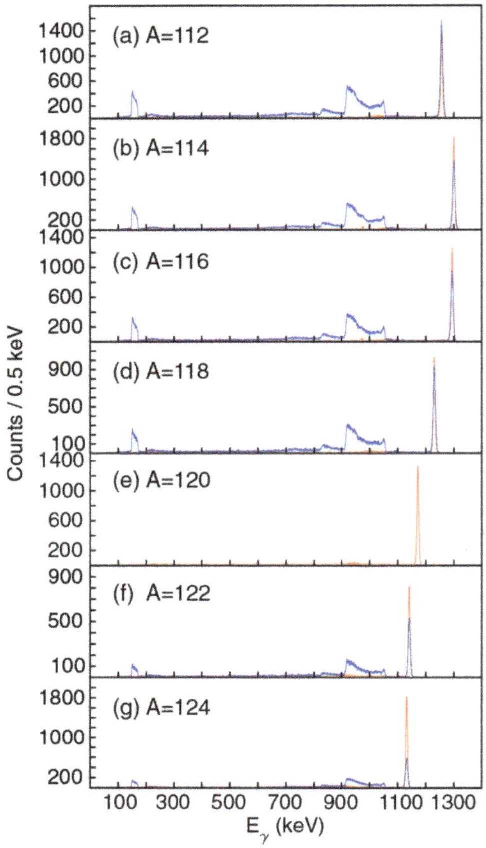

Figure 6.2. Coulomb excitation of the nuclei $^{112-124}$Sn. These data are for beams of the respective Sn isotopes on targets of natural C and natural Ti (the red and blue spectra are for the C and Ti targets, respectively). The γ-ray spectra show the population of the first excited 2^+ states in these isotopes (cf. figure 2.13 and 4.5). The Doppler broadened γ-ray lines corresponding to the first-excited states in the Ti isotopes are also visible (see text for details). The γ-ray yields can be directly converted to $B(E2; 2_1^+ \rightarrow 0_1^+)$ values. Note, the very similar energies of these 2_1^+ states: by using beams of the nuclei of interest, the accelerator functions as a very high-resolution mass spectrometer. If the nuclei of interest were the targets, there would be isotopic impurities in each data set because 100% pure isotopic samples are impossible to achieve. Reprinted with permission from [2]. Copyright (2015) by the American Physical Society.

Conversion of γ-ray yields to $E2$ (and $E3$) matrix elements is the challenge in the data analysis. Details of this lie beyond the present level of treatment, but will feature later in the Series.

We note that $E1$ excitation is not observed at low energy in nuclei. The reason is that nuclei do not possess electric dipole moments and they do not possess low energy vibrations with electric dipole character. In simple terms, the centre of mass and the centre of charge in a given nucleus would have to separate with low 'rigidity' to such separation. Nuclei only possess a highly rigid $E1$ excitation, the so-called giant electric dipole resonance which is observed at >10 MeV in nuclei.

A subtle point in Coulomb excitation is that parity-changing excitation occurs through $E3$ multipoles; but parity-changing de-excitation, i.e. radiative decay occurs (predominantly) through $E1$ multipoles. The decay probabilities reflect the Weisskopf estimates for electromagnetic radiation, i.e. the coupling of the nucleus to the electromagnetic vacuum. The excitation probabilities reflect the interaction between the electrical charges of two nuclei in relative motion (target and projectile) coupling via the electromagnetic interaction.

6.3 Radioactive decay

An easy way to study excited states in nuclei is via radioactive decay of a parent nucleus to the daughter, the latter being the nucleus of primary interest. The advantages are that the parent may have a long half-life and so studies can be conducted remotely from production facilities. With so-called tape-transport systems, even short-lived activities can be moved rapidly from production point to customized detector arrangements. Further, the technology for on-line isotope separation is highly advanced with respect to selection of mass and, via selective ionization, selection of Z.

The use of spectroscopic measurements following radioactive decay has been employed since the times of Ernest Rutherford. It is the 'time-honored' technique in nuclear structure study. By far the prevalent source of such data is β-decay scheme studies, β^+/EC and β^- decays. A schematic view of such decay, as a function of spin and excitation energy, is shown in figure 6.3. There are stringent spin selection rules for β decay[1]—usually this is limited to $\Delta I = 0, \pm 1$. This has the advantage that assignments of spin values are constrained. It has the disadvantage that only low-spin states can be studied. There is the natural limit that population of excited states does not exceed the beta decay energy, Q_β.

Elucidation of a radioactive decay scheme is, in general, a complex task. But a given decay scheme is uniquely reproducible, i.e. the scheme is an invariant aspect of the given parent nucleus, which obeys universal laws of quantum probability under all laboratory conditions. Thus, radioactive sources serve as excellent standards for calibration of γ-ray energies and intensities. Some details are given in chapter 7 under measurement standards. Some details of the spectroscopic techniques for assigning spins and parities to excited states in the daughter nucleus are presented in chapter 2, notably the use of conversion electron spectroscopy and angular correlations between coincident pairs of γ-rays. Some more details are presented shortly.

Transitions occurring between states in the parent nucleus and the daughter nucleus provide information on transition matrix elements for β-decay and α-decay. The β-decay process is controlled by the weak interaction. The α-decay process is controlled by penetration of a Coulomb barrier by the α particle. These matrix

[1] The spin selection rules for β decay are dominated by the two possible spin couplings for the electron (positron) and antineutrino (neutrino): the emission of the spin-1/2 pairs of particles can be in $S = 0$ or $S = 1$ correlated states, termed 'Fermi' and 'Gamow–Teller' β decay, respectively. With lower probability, orbital angular momentum can also be removed from the β decaying parent nucleus by the emitted pair.

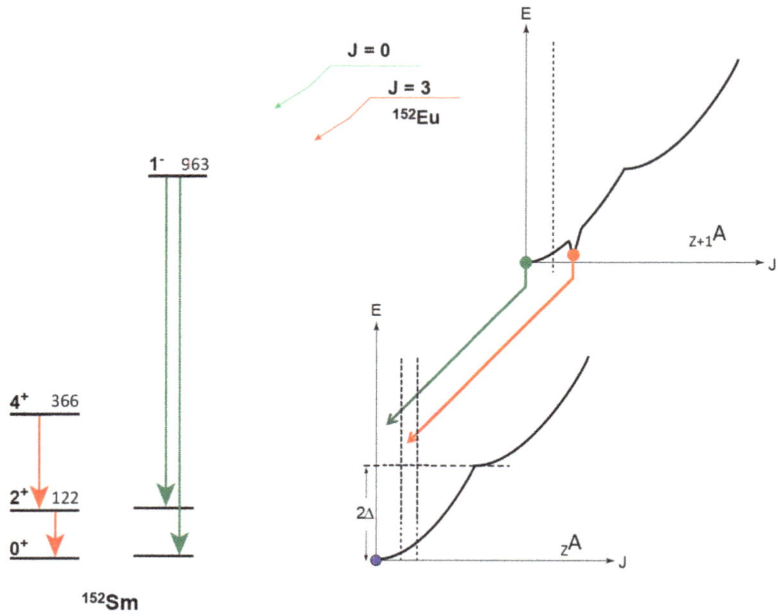

Figure 6.3. A schematic view of a β^+-decay scheme is shown. It involves the decay of the ground state and a low-lying isomeric state in the nucleus ^{152}Eu to the ^{152} Sm daughter. The spin-selection rules for β-decay favor $\Delta I = 0$ or 1. The entire image is set in the framework of spin versus excitation energy to emphasize the favoring of low-spin population of daughter states in β decay. Note that in this particular pair of decays, the low-spin β-decaying state in ^{152}Eu is an excited state at 45.6 keV, not the ground state (contrary to the schematic view on the right). In a few rare instances, β decay from high-spin states is observed (cf. figure 3.22). The energy '2Δ' represents the pairing energy gap, above which the level density becomes high (cf. figure 2.13 and 3.16).

elements can provide information on the structure of the parent ground state and on the directly populated daughter states. However, a word of caution is needed with respect to complex β-decay schemes: in cases where the β-decay energy is large, there may be many weak and unobserved decay branches. Thus, many reported β-decay branches to states in daughter nuclei must be viewed with caution (even disbelief). This pathology in nuclear spectroscopy is sometimes referred to as the 'pandemonium' effect [3].

Beyond pandemonium effects, another source of common error in decay scheme studies is the failure to identify sum peaks. These occur between pairs of γ-rays in prompt cascades: by 'prompt' is meant no time delays due to long-lived intermediate states. 'Long-lived' is somewhat subjective, depending on coincidence timing, but $>$50 ns would be considered 'long' and $<$10 ns would be considered 'prompt'. With conversion-electron spectroscopy, it is important to keep in mind that, e.g. internal conversion of a K shell electron is most probably followed by the prompt emission of a K x-ray (less probably an Auger electron). Thus, summing between K-shell conversion electrons and K x-rays must be quantified when detectors sensitive to both types of radiation are being used. (This is avoided by conducting electron spectroscopy with magnetic spectrometers or filters.) Detection efficiency has a $[\text{distance}]^{-2}$ dependence for single γ-rays, so the occurrence of summing has a [distance]$^{-4}$ dependence, i.e. the relative incidence of summing can be reduced, by increasing the source-to-detector distance.

A further aspect of β decay is that for β^- decays with Q_β values that exceed the one-neutron separation energy in the daughter nucleus, β-delayed neutron emission[2] can occur. This complicates the decay scheme in that two mass chains become involved. However, this is also a means of studying excited states in both the β-decay daughter and the β(n)-decay daughter. For β^+ decays with Q_β values that exceed the one-proton separation energy in the daughter nucleus, a similar β(p)-decay can occur, but there is a hindrance due to the Coulomb barrier for the proton with respect to emission from the β-decay daughter nucleus.

6.4 In-beam reaction γ-ray spectroscopy

In-beam spectroscopy had an obvious beginning with the recognition that Coulomb excitation could be used. It had the less obvious beginning that prompt radiations associated with nuclear reactions could be used. The latter realization was by Morinaga and Gugelot, working at the IKO in Amsterdam [4, 5]; and major pioneers were Frank Stephens and Dick Diamond, using heavy-ion induced reactions conducted at the Lawrence Berkeley Laboratory [6]. Generally, in-beam spectroscopy populates high-spin states.

There was a major revolution for in-beam spectroscopy with the recognition that arrays of Compton-suppressed detectors provide an enormous advance in sensitivity. This development was due to the nuclear spectroscopy group at the University of Liverpool. It was conducted at the Daresbury Accelerator Lab (now closed). Experiments revealed the widespread occurrence of excited bands in nuclei that are so-called 'superdeformed', i.e, far more deformed than any observed ground state deformed structures [7].

Superdeformed band exploration dominated the field of low-energy nuclear structure study for some 30 years. In essence, this was a completely 'new face' for nuclear structure. However, the experiments to determine key moments of the structures, even their absolute excitation energies and spin-parities, proved incredibly demanding. Very recently, a dedicated study of the decay of a superdeformed structure has revealed that the 'draining' of these superdeformed structures is extraordinarily complex and involves many hundreds of transitions [8]. At present, it is not possible to present a systematic view of superdeformation in nuclei without relying on a heavy role for model-based interpretations.

The main reason that the study of excited states in nuclei by in-beam reaction γ-ray spectroscopy has been an eminent sub-field of nuclear spectroscopy is that bands of states with monotonically increasing spins dominate nearly all of nuclear structure, and these are preferentially populated by reactions. Thus, such spectroscopy is a powerful means of establishing excitation patterns in nuclei. Further, by

[2] We note that β-delayed neutron emission is critical to the stability of nuclear reactor operations: it provides a source of neutrons for sustaining the chain reaction that is 'damped' by the (short) β-decay half-lives of the prompt fission products.

judicious choice of target, projectile and projectile energy, enhancement of the nuclei of interest for study can be achieved. Yet further, by detecting recoiling nuclear species—by mass and charge, evaporated protons and neutrons, even ground-state decay events of the nucleus of interest, following the reactions occurring (often running to tens of diverse reaction 'channels')—in coincidence with γ-rays, remarkable selectivity of extremely rare reaction products can be studied. Spin selectivity for state population can be made by the choice of mass and energy of the projectile. An example of excited states determined by the ^{92}Mo(^{88}Sr,p2n)^{177}Au reaction, with α-decay 'tagging', is shown in figure 6.4.

However, a 'gap' in spectroscopic detail occurs as one moves away from the stable nuclei. To reach nuclei far from the stability region requires heavy-ion reactions and such reactions predominantly populate high-spin states, while radioactive decay populates generally low-spin states: the result is a severe lack of information with respect to states of intermediate spin. We note that, when using beams of charged particles, to produce a nuclear reaction (leading to the desired species of study), the beam energy must be sufficient to overcome the Coulomb barrier in order for a nuclear reaction to occur. Higher beam energies will impart higher angular momentum to the product nucleus. (Also, higher beam energies lead to more 'evaporation' channels from the compound nucleus, i.e. the nucleus formed

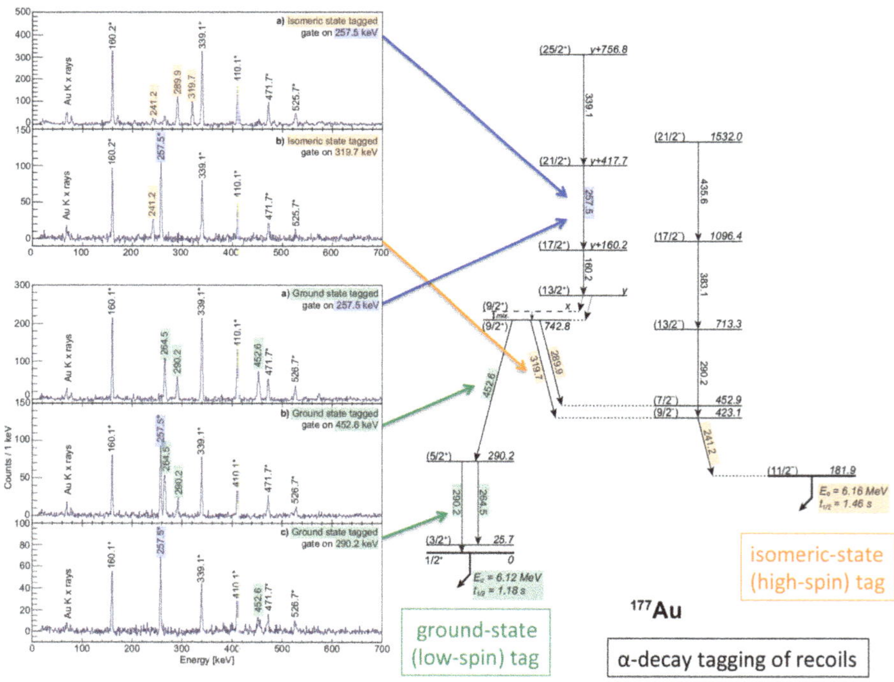

Figure 6.4. An illustration of population of bands of states in ^{177}Au by the ^{92}Mo(^{88}Sr,p2n)^{177}Au fusion–evaporation reaction. The bands are assigned to the nucleus ^{177}Au by tagging γ-ray events using α decays of the recoiling ^{177}Au nuclei. This is highly selective for ^{177}Au, and reveals a complex decay pattern, which is directly inferred from the gated spectra as explicitly shown. The figure was supplied courtesy of M Venhart and is similar to one appearing in [9].

by the instantaneous fusion of the projectile nucleus and the target nucleus.) Even with long running times to observe weakly populated bands by in-beam spectroscopy, limited accelerator beam-time availability has resulted in major gaps in systematic spectroscopic detail. We reiterate that the study of the nuclear many-body problem demands systematic study of nuclei as a function of Z and N, and as a function of spin.

One singularly successful type of in-beam reaction γ-ray spectroscopy is provided by inelastic scattering of beams of mono-energetic neutrons. Varying the neutron beam energy, it is possible to stepwise progress to higher and higher excitation energy in the nucleus under study. The excitation process is exclusively through the strong interaction and is found to be non-selective. For a given beam energy, there is manifestly an energy of excitation limit; there is also a spin limit. In consequence an excitation 'map', for a selected nucleus, can be comprehensively explored: in principle, such exploration can be pursued to a 'complete' spectroscopic picture of the nucleus studied (up to a given energy and spin). However, there are significant limitations: large quantities of highly enriched isotopes are needed (this is expensive, even when leased) and the technique is only applicable to stable isotopes; further, large source mass attenuates low-energy γ-ray detectability. But, the technique has been developed, via quantification of angular distributions of γ-rays, excitation functions for observation of γ-rays as a function of neutron-beam energy, and Doppler energy shifts in γ-ray energies, to yield transition multipolarities, level lifetimes and hence γ-decay matrix elements, and spins of decaying states. A sample of such broad spectroscopic detail is depicted in figure 6.5. Figure 6.6 shows the progressive excitation of states as the neutron bombarding energy is step-wise increased. Such a program is pursued at University of Kentucky, which operates a unique program of such measurements.

A source of most-valuable data for nuclei is prompt γ-ray emission following nuclear fission. Such a process provides information on neutron-rich nuclei. Further, the average spin of the states populated in the fission fragments is $\sim 10\hbar$, with a 'reach' up to $\sim 16\hbar$. Fission can be spontaneous as occurs in the decays of ^{252}Cf ($T_{1/2} = 2.6$ y, 3.1% fission decay branch) and ^{248}Cm ($T_{1/2} = 3.5 \times 10^5$ y, 8.4% fission decay branch), or it can be induced by reactions on, e.g. a uranium target. What is critical in such studies is the role of large arrays of Ge detectors with the capability for triple-coincidence detection between cascades of γ-rays. The enormous variety of fission products that are involved in the statistical distribution require 'sorting' into isotopic species via the knowledge of the energies of at least two correlated events. This correlated pair may be in a single nucleus or in a pair of fission partner products (with allowance for the loss of a few neutrons by 'evaporation'). Details are beyond the present level of discussion.

A type of reaction spectrosocpy that has a unique standing in nuclear structure study is co-called 'capture γ-ray' spectroscopy. (This extends to capture conversion-electron and internal-pair spectroscopy.) This involves processes such as neutron capture or proton capture on (target) nuclei. The capture may be into specific unbound resonances. In the case of neutron capture, often 'thermal' neutrons are employed, i.e. neutrons that have been slowed in a reactor core to speeds corresponding to 300 K

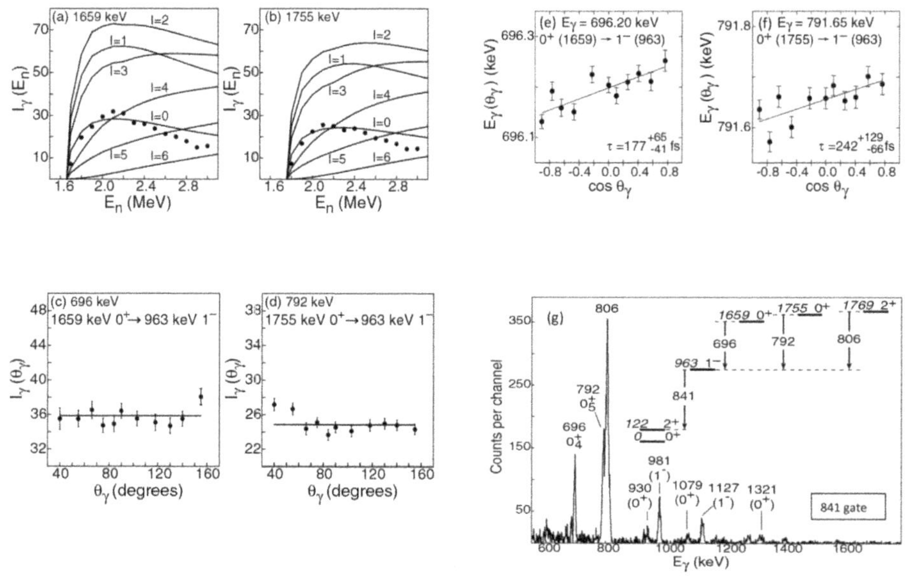

Figure 6.5. Spectroscopic data for ^{152}Sm obtained by inelastic scattering of neutrons. The neutron beams were produced by bombarding a tritium target with protons, with variation of the proton beam energy in 0.1 MeV steps, so as to yield $E_n = 1.2, 1.3, 1.4, \ldots, 3.0$ MeV. The spread in neutron energy for each step was ~0.1 MeV. This permitted sequential excitation of the states in the ^{152}Sm target nucleus, yielding excitation functions for de-exciting γ-rays—(a) and (b), which provide threshold of excitation information and spin information for the states. Recording γ-ray spectra as a function of angle with respect to the beam direction provides further spin information, as shown in (c) and (d) (the 'flat' or isotropic response shows that the 696 and 792 keV γ-rays originate from levels with spin zero). These data, as a function of angle exhibit small but precisely measurable Doppler energy shifts that can be converted to lifetimes of excited states—(e) and (f). The result is information with high-spectroscopic detail for complex levels schemes such as exhibited by ^{152}Sm: summarized in the sub-scheme shown in (g); the γ-ray spectrum is a coincidence gate on the 841 keV transition (cf. figures 2.1, 2.2, 6.1 and 6.6). Reprinted with permission from [10]. Copyright (2008) by the American Physical Society.

(= 2200 m s^{-1}). Also, sometimes filtered beams of fast reactor neutrons are used, whereby capture is averaged over many resonances. For proton capture, the proton beam energy can be tuned so that capture occurs in so-called isobaric analog resonances. Overall, capture γ-ray spectroscopy is very complex; but it can provide a comprehensive view of low-spin excited states in nuclei that can be reached by addition of one proton or one neutron to a stable (target) nucleus. Details will be taken up later in the Series.

6.5 Gamma-ray spectroscopy

Generally, γ-ray spectroscopy is conducted with germanium detectors. The technical details of the different types are beyond the present level of discussion. However, there is now frequent use of so-called 'Compton suppression' whereby the core Ge detector is surrounded by a separate annular detector which vetoes any event that involved a coincidence between the core detector and the annular detector because

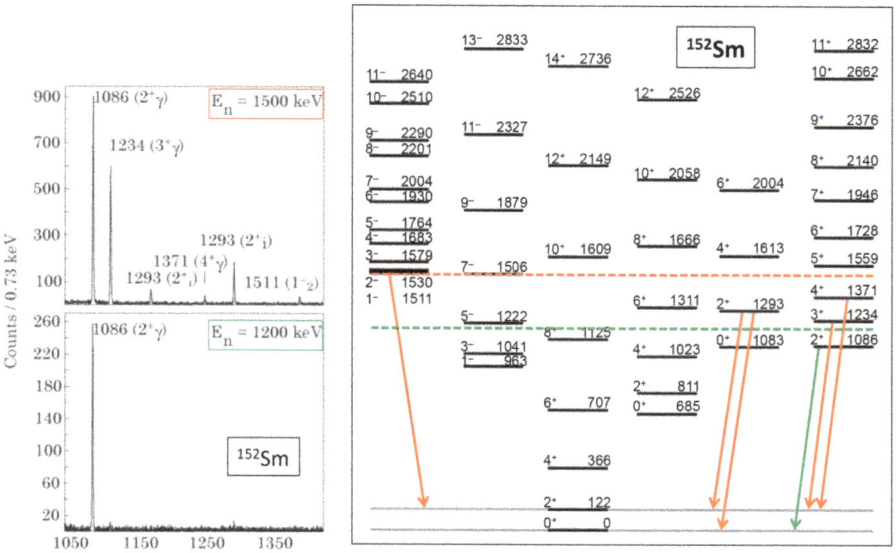

Figure 6.6. Two steps in sequential excitation of states in ^{152}Sm using the $(n, n'\gamma)$ reaction. The neutron energies are E_n = 1200 and 1500 keV. The energy range shown for the γ-rays is 1050–1400 keV; thus, a number of levels that are populated are not evident in the spectrum. Note that the γ-ray lines are labeled by the energies of the states from which they originate. Note also that the population of the level at 1511 keV is because the neutron beam energy at 1500 keV has a spread. The figure is adapted from one that was supplied courtesy of W D Kulp.

such an event is likely due to Compton scattering of a γ-ray. If this veto is not made, the core detector records partial energy events as well as full energy events and this leads to a spectrum background that obscures weak peaks. Recall, the 'statistics' of radiation spectra is Poisson statistics and so for N events in a spectrum channel, there is an inherent fluctuation uncertainty of $\pm\sqrt{N}$. A good Compton suppression system can reduce the background counts by ~3 × and so a $\sqrt{3}$ = 1.7 increase in statistical sensitivity is achieved.

Coincidences between pairs of γ-rays generally increase sensitivity by significant amounts for identifying weak decay branches. If coincidence intensities are needed, a few cautionary points must be considered. First, there are angular correlations between pairs of cascade γ-rays, which can result in considerable attenuation of observed intensities. Second, there is a subtlety in the coincidence relationship between the 'feeding' γ-ray and the 'draining' γ-ray with respect to a level. This is quantified in figure 6.7. Below some (low) energy threshold, processing of events by the electronics will result in losses: this must be carefully determined with calibration sources.

One capability that is exceptional for radioactive decay schemes is that they are well suited to the determination of extremely weak decay branches. An example of highly critical importance for structure determination is illustrated schematically in figure 6.7. Such a measurement necessitates acquisition of coincidence data. Very weak, low-energy transitions between higher lying excited states are the critical proof of collective band structures. An example is presented in figure 6.8.

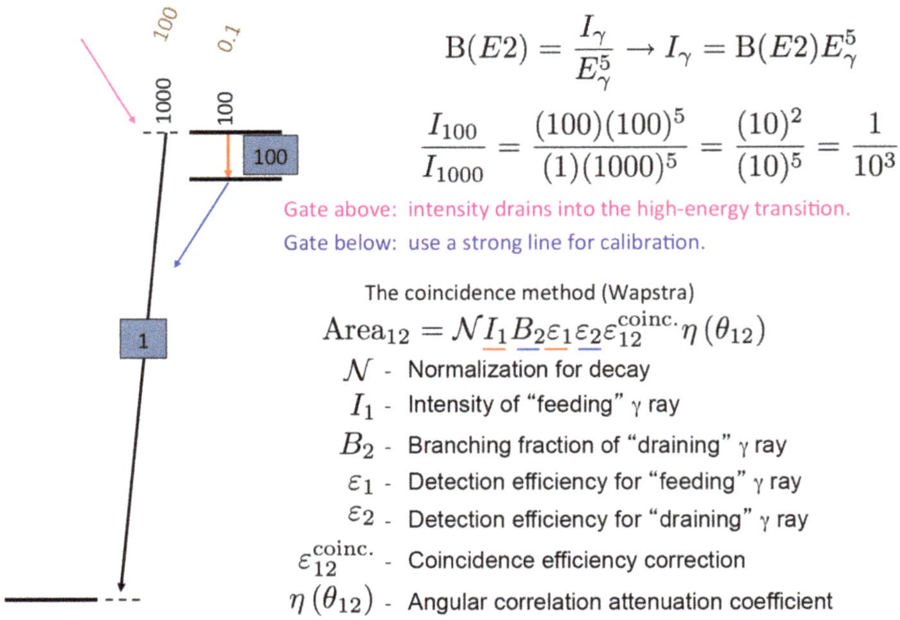

Figure 6.7. A fundamental problem with respect to direct proof of a band structure is the direct observation of very weak, low-energy decay branches between higher lying levels. The relevant equations are presented in the figure. Radiative decay by *E*2 transitions possesses an E_γ^5 factor: this has the result that low-energy transitions in competition with high-energy transitions are frequently not observed. To optimize direct observation of such low-energy transitions, coincidence gating should be 'from below', i.e. gating on a draining transition following the low-energy transition, not a feeding transition preceding the low-energy transition. An example is shown in figure 6.8. The figure is adapted from one that was supplied courtesy of W D Kulp.

We reiterate that radioactive decay is a 'passive' population of nuclear excited states (it does not involve the active role of an accelerator); as such, very long data acquisition times can be used to observe such very rare decay branches, if the half-life of the parent isotope is long.

Coincidences between γ-rays are the basis of γ–γ angular correlations. Some examples are presented in figure 2.3. Such data provide information on the spins of the initial, intermediate, and final states and on the multipolarities of the intervening transitions. In particular, a commonly occurring mixed multipolarity is $E2 + M1$; such transitions are possible when the spin change is 0 or 1, there is no parity change (and the spins of the initial and final states of each transition are not 0 and 0 or 0 and 1). The angular correlation pattern, $W(\theta)$, such as shown in figure 2.3, can be fitted with an expression of the form

$$W(\theta) = a_2 P_2(\cos\theta) + a_4 P_4(\cos\theta), \tag{6.1}$$

where a_2 and a_4 are fitted to the data and the $P_L(\cos\theta)$ are Legendre polynomials. An example is shown in figure 6.9. We note that sometimes $E2 + M1$ transitions are very strongly dominated either by the *E*2 component or by the *M*1 component. While conversion electron spectroscopy in principle can quantify $E2 + M1$ admixtures, in practice only γ–γ angular correlations can reliably do this when

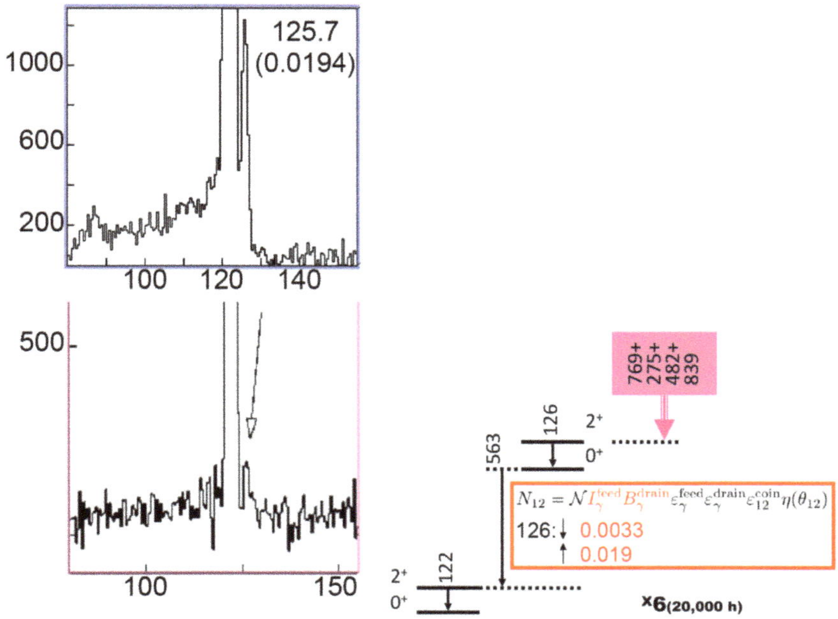

Figure 6.8. Illustration of the difference between coincidence gating: 'from above' versus 'from below'. The data are for the β^+/EC decay of the ^{152}Eu ground state to ^{152}Sm. The upper spectrum shows gating from below for 252 h of data acquisition to observe the 125.7 keV transition (between the 811 and 685 keV levels in figure 6.6), the lower spectrum shows summed gates from above for 600 h of data acquisition (with a comparable detector array). There is a factor of 6 enhancement for gating from below, this translates into a factor of 36 enhancement for Poisson statistics, i.e. gating from above would necessitate over 20 000 h of data acquisition time to achieve the peak-to-background ratio achieved in the upper spectrum. The 125.7 keV transition has a branching fraction of 5.1×10^{-5} per ^{152}Eu decay. The spectra are taken from [11]—upper spectrum, and [12]—lower spectrum, and are reprinted and cropped with permission. Copyright (1999 and 2007) by the American Physical Society.

one or other multipolarity is dominant. Knowledge of $E2$ admixtures in such transitions is essential for elucidating patterns of nuclear collectivity.

We note that there have been a few studies of angular distributions of γ-rays emitted from 'oriented' nuclei, i.e. the spins of β-decaying parent nuclei are polarized through their magnetic moments in a powerful magnetic field. The field that produces the orientation is the internal magnetic field in an Fe foil: this is polarized by application of a magnetic field of ~0.2 T. This must be conducted at 'low temperature', i.e, ~10 mK, which requires a ^{3}He/^{4}He dilution refrigerator. Sometimes nuclear magnetic resonance is coupled to the measurement samples and anisotropies of radiations are detected as the RF is modulated to identify the resonances. A few such refrigerators have even been operated 'on-line' to study short-lived radioactive isotopes. A subtle point is that the temperature of 10 mK must be maintained with an exposure (through the implantation aperture) to 300 K: this requires 'heat pumping' power at 10 mK. It is important to note that the Ge detectors used must be deployed near a powerful magnetic field: not all detector types are suited to such an environment.

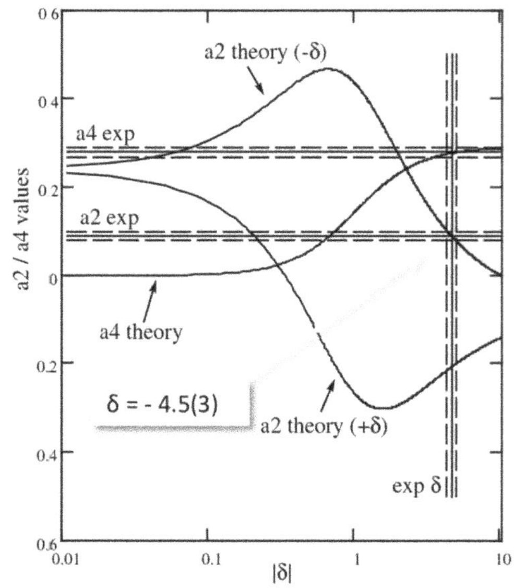

Figure 6.9. The determination of an $E2/M1$ multipole mixing ratio by γ–γ angular correlation measurements. This is shown for a $2^+\rightarrow2^+$ transition (the $2_2^+ \rightarrow 2_1^+$ transition in ^{150}Sm, populated in the β^- decay of ^{150}Pm using the $2_2^+ \rightarrow 2_1^+ \rightarrow 0_1^+$ cascade). The angular correlation pattern, as shown, e.g. in figure 2.3, is fitted with a function of the form equation (6.1) to yield a_2 and a_4 coefficients. These coefficients provide a precise value for $\delta(E2/M1)$. Note the importance of the sign of this quantity: often the a_4 term is lacking in such measurements and then two possible values of δ are quoted. The figure is taken from the PhD dissertation of P Schmelzenbach (Oregon State University, 2003) and is reproduced with permission. The figure is similar to one appearing in [13].

The interaction between a γ-ray and a detector can occur via: photoelectric events, whereby all of the energy of the γ-ray is deposited in the detector by one or more photoelectric interactions; by Compton scattering, whereby only part of the photon energy is imparted to the detector (but the total will be recorded if the Compton scattered photon then undergoes a photoelectric event); and by electron–positron pair formation. The ranges of electrons in the detector material are small and so there are negligible losses if all of the energy of the photon is imparted to electrons. But positrons eventually annihilate, yielding two 511 keV photons: one or both of these photons may not deposit their energy in the detector, with the result that one observes so-called 'single-escape' and 'double-escape' peaks, these occur at E_γ – 511 keV and E_γ – 1022 keV. The threshold for this latter process is for γ-rays with energies above $2 \times 511 = 1022$ keV, with significant probability that this process occurs above 2 MeV. There is also a very small probability for Ge x-ray escape peaks being observed, at E_γ – 9.9 keV. The ionization produced in the detector is collected and amplified to yield a pulse with voltage amplitude that is directly and accurately proportional to the energy deposited by the γ-ray.

The efficiency curve for γ-ray photon detection as a function of γ-ray energy reflects the composite response of the detector to photoelectric interaction, Compton scattering, and electron–positron pair formation. While these can be complex, the integrated detection efficiency conforms to a standard profile. Above the energy where attenuation of low-energy photons prevents all of them from reaching the active volume of the detector, efficiency curves closely obey a relationship of the form

$$\epsilon = AE_\gamma^{-r}, \tag{6.2}$$

where ϵ is the detector efficiency and A and r are constants. This holds to within 5%, typically from ~200 keV up to ~1500 keV. The constant A depends on factors such as source-to-detector distance. The constant r takes values between ~0.7 for 'large volume' detectors and ~1.0 for 'small volume' detectors, for typical 'off-the-shelf' detectors. Thus, by tabulating $A = \epsilon E_\gamma^r$ it is possible to proceed to high-precision efficiency curves via fitting deviations from this relationship. This can assist interpolation between multiple calibration sources.

Gamma-ray detector technology advances at a steady pace. Figure 6.10(a) and (b) shows a recent advance in detector resolution and line shape that is creating a revolution in the elucidation of complex decay schemes. Beyond the obvious advantage of higher energy resolution, the BEGe detectors[3] involved have near-perfect Gaussian line shapes that provide a major advance in de-convoluting γ-ray 'multiplets' (which are always present, no matter how high the energy resolution). Further, arrays of 'segmented' Ge detector units are now beginning to be employed. Such arrays 'track' any Compton scattering processes and record such events.

6.6 Electron spectroscopy

Electron spectroscopy covers internal-conversion electrons and internal pairs (pair = electron–positron pair, IPs). All γ-ray transitions in nuclei compete with the internal conversion (IC) process: some details appear in chapter 2. Above a transition energy of 1022 keV, the IP process also competes. These processes variously have probabilities that depend very strongly on: transition energy, transition multipolarity, and atomic number (Z). Particularly, parity changes in nuclear transitions are easily identified using IC electron studies. Further, there is one nuclear decay process that can only be detected by IC electron spectroscopy, and above 1022 keV by IP spectroscopy: the occurrence of electric monopole ($E0$) transitions.

Two basic types of spectrometer are in use for electron and positron spectroscopy: semiconductor detectors and magnetic spectrometers. (There have been a few studies with spectrometers using electrostatic fields.) Indeed, prior to the advent of semiconductor detectors in the period 1963–65, magnetic spectrometers were used directly for high energy-resolution spectroscopic studies of nuclei[4]. This was carried

[3] The term 'BEGe = broad energy range germanium' is simply one of naming convenience and has its origin in the property that such germanium detectors possess very high energy resolution over an unprecedentedly broad energy range. For details of performance characteristics of such a detector, as applied to nuclear structure study, see [14].

[4] Pioneering nuclear spectroscopic work using magnetic spectrometers was initiated at the Univeristy of Uppsala in the 1950s based on the historical contributions to x-ray spectroscopy by Manne Siegbahn (Nobel Prize 1925) and by Kai Siegbahn's contributions to high-resolution electron spectroscopy (Nobel Prize 1981, shared with Nicolaas Bloembergen and Arthur Schawlow for their respective contributions to laser spectroscopy).

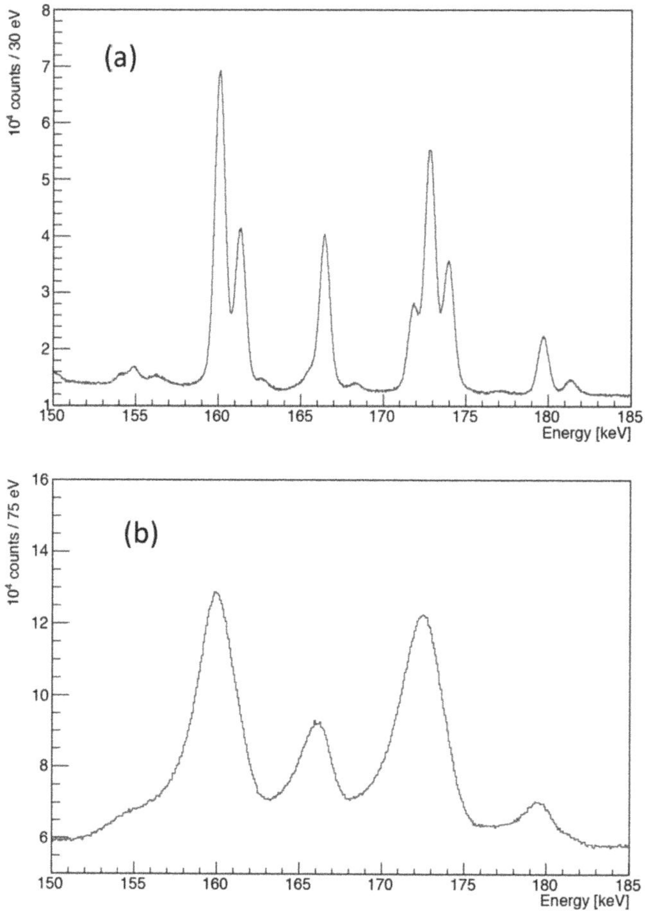

Figure 6.10. Comparison of γ-ray spectra acquired with (a) a so-called BEGe detector and (b) a standard coaxial p-type Ge detector. The energy resolution of the p-type Ge detector is 2.0 keV for the 166 keV singlet peak and 680 eV for the BEGe detector. Beyond the obvious resolution of doublet and triplet peaks, a more subtle issue is the line shape difference. The p-type Ge detector lines have a low-energy 'skew', whereas the BEGe detector lines appear (and are near perfect) Gaussian in shape. Note the very high 'gain' of 27 eV/ch, which permitted a high-resolution view of the BEGe line shapes. To computer-fit multiple peaks in a multiplet is greatly facilitated when the shapes are not skewed. Here, the spectra involve parent and daughter decays in the A = 183 isobars. It is possible with precise peak fitting to achieve a very high degree of de-convolution of parent and daughter decays. While differences in half-lives of the parent and daughter decays helps in such a deconvolution, it is impossible to obtain mono-isotopic spectra in such an A-chain sequence of decays and so de-convolution is unavoidable. (Note: such detectors can be expected to lose Gaussian line shape if there is a neutron background such as with in-beam spectroscopy.) The spectra are supplied courtesy of M Venhart. More details are presented in [14].

out directly by observation of the internal-conversion electrons, and indirectly via the emitted γ-rays impinging on an 'external converter', e.g. a tungsten foil which produced photoelectrons at an energy of $E_\gamma - B_K$ (69.529 keV). Hence, we use the terms 'external-conversion' electrons and 'internal-conversion' electrons. We would urge our Reader to familiarize themselves with this historical period in nuclear spectroscopy: many basic structural features, such as spins and parities of isomeric states in nuclei, were elucidated in these early studies.

Electron spectroscopy must be conducted under a high vacuum with 'open' sources, i.e. no protective covering, otherwise the electron energies are degraded. In consequence, conversion-electron spectroscopy has received only limited attention by the nuclear physics community. This has been particularly so with the move to studies of nuclear structure using short-lived radioactive species. In-beam IC electron study has its own set of challenges, notably so-called delta electrons, best described as low-energy electrons 'splashed' out of the target by the impinging beam, and hitting the detector. Magnetic lenses can filter out delta electrons. However, magnetic analysis precludes acquisition of coincidences between IC electrons and γ-rays (or other IC electrons).

Internal-pair spectroscopy necessitates detection and recording the total energy of the electron plus the positron. This type of spectroscopy has only been pursued by a handful of practitioners in the entirety of the history of nuclear spectroscopy. There is now a dedicated device for such work located at Australian National University (ANU), cf. figure 6.11. We reiterate, IC electron and IP spectroscopy are the only means to study $E0$ transitions in nuclei. As described in chapter 4, such transitions play a crucial role in the study of shape coexistence in nuclei. In light nuclei, IP spectroscopy is the only viable means of investigating the occurrence of the $E0$ decay process. Recently, the ANU Group reported the $E0$ decay branch from the Hoyle state in ^{12}C [15]: in all of us, a significant number of carbon atoms arrived in our bodies, from the primordial hydrogen and helium of the Universe, by way of this decay path.

A further unique aspect of conversion electron spectroscopy is that very low energy transitions in nuclei are 'highly converted', i.e. the transitions occur almost entirely by the internal conversion process: therefore, such transitions can only be observed using electron spectroscopy. Such spectroscopy is extremely difficult to perform, primarily because very low energy electrons are absorbed by the actual radioactive source, unless the source is 'mass free', i.e. is near to an atomic monolayer in thickness. An example of such a spectrum is shown in figure 6.12. (We note that mass-free sources can now be prepared in a trap, see [16].) Electron spectroscopy is effectively carried out using lithium-drifted silicon, Si(Li) detectors. A major advantage of such detectors is that they can be coupled with Ge detectors, or other Si(Li) detectors in coincidence mode. This becomes essential for complex decay schemes.

6.7 Reaction spectroscopic studies of nuclear structure

Reaction spectroscopy refers to measurement of the energy of an outgoing particle (or sometimes a pair of particles) following a nuclear reaction. Particularly, this applies to inelastic scattering of, e.g. protons, deuterons and alpha particles (inelastic electron scattering is addressed shortly). It also applies, particularly, to one- and two-nucleon transfer reactions, e.g. single-neutron transfer—(p, d), (d, p), (d, t),

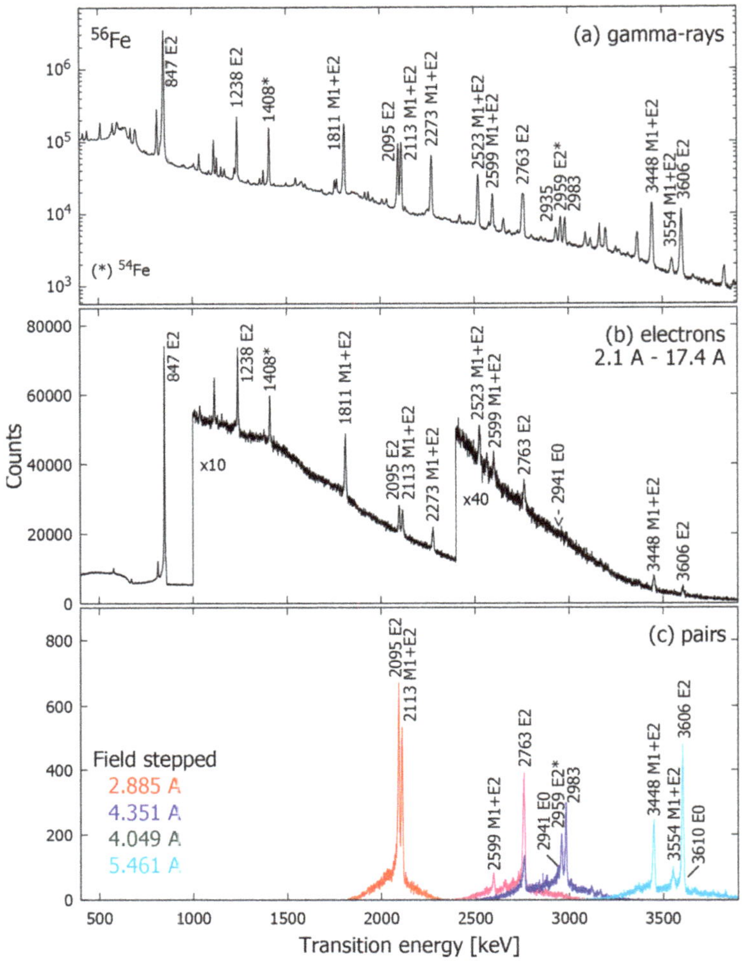

Figure 6.11. A comparison of γ-rays, conversion electrons, and internal pairs: from the excitation of ^{56}Fe by inelastic scattering of protons. The spectra are presented so that *K* conversion lines and internal-pair lines are matched. The *K*-electron binding energy of iron is 7.1 keV. The internal-pair lines are shown, shifted by 1022 keV. Note that the internal-pair spectrum is a composite of four separate settings of magnetic spectrometer current (given in amps). The figure was supplied courtesy of T Kibédi.

(^{3}He, α), (α, ^{3}He); single-proton transfer—(^{3}He, d), (α, t), (t, α); two-neutron transfer —(p, t), (t,p); and via time-of-flight measurements, two-proton transfer—(^{3}He, n). Multi-nucleon transfer reactions also provide nuclear structure information and examples include (p, α), (d, ^{6}Li), (^{7}Li, t), (^{16}O, ^{14}C), and various other combinations. The primary spectroscopic necessities are a mono-energetic beam and a high-resolution magnetic spectrograph. Such studies even extend to polarized beams. A schematic view of the experimental arrangement is shown in figure 6.13.

Data reduction depends upon a reliable description, with respect to the target nucleus, of the interaction with the incoming projectile nucleus ('entrance channel' process) and of the interaction with the outgoing ejectile nucleus ('exit channel'

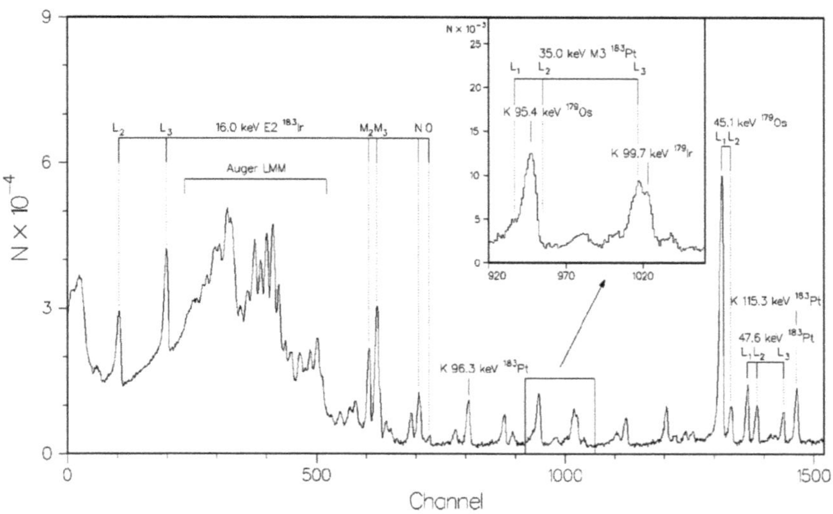

Figure 6.12. An electron spectrum: which shows details at ultra-low energy. The lowest energy electron line is at 3.2 keV. The LMM Auger electron lines are evident, i.e. the filling of an L-shell vacancy by an M shell electron, with the emission of a second M-shell electron that acquired the energy released in the process (there are many possible combinations, hence the complex multiplet). The spectrum was acquired with a magnetic spectrometer. Reprinted from [17] with permission from Springer-Nature, copyright (2000).

process). This is formulated using so-called 'optical' models. Allowance must be made for Coulomb excitation in the entrance channel and in the exit channel. These are termed 'coupled channel' effects. Details are beyond the present level of discussion.

Inelastic scattering of charged particles provides some information that is similar to that obtained by Coulomb excitation. But it possesses another capability. Information on the isospin structure of nuclear transitions can be obtained. Thus, e.g. inelastic scattering of deuterons and alpha particles probe so-called isoscalar excitations; whereas inelastic scattering of, e.g. protons and tritons probe isovector excitations. There have even been limited studies of inelastic scattering of pions, both π^+ and π^-, which from their quark–antiquark composition, π^+—$u\bar{d}$, π^-—$\bar{u}d$, exhibit extreme sensitivity to the proton (uud) and neutron (udd) contributions to nuclear transition processes (anti-quarks do not experience a Pauli exclusion principle in nuclei).

Inelastic scattering can also be used to probe nuclear transition processes as a function of nuclear radius. This is done by mapping so-called 'form factors', which is done by measuring the probability of the transition of interest as a function of beam energy and scattering angle. These are more advanced topics and are presented later in the Series.

A critical ingredient of reaction spectroscopy is energy resolution: the accelerator beam must be mono-energetic (as nearly as possible) and the magnetic analysis of the ejectile nuclei must be of the highest attainable quality. Current world-best energy resolution is ~3 keV. It is regrettable that in current times there are very few facilities that offer high-resolution reaction spectroscopy capabilities. Until very recently, the tandem accelerator laboratory at Ludwig Maximillian University in

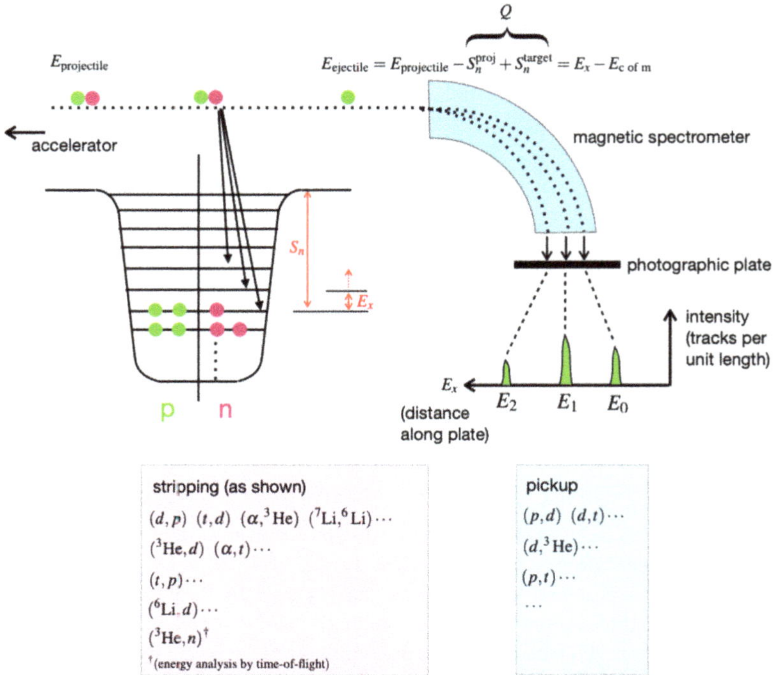

Figure 6.13. A schematic view of the experimental arrangement of a magnetic spectrograph used to conduct reaction spectroscopy. The detail shows how a transfer reaction such as (d, p) populates states in a shell model potential. Such spectrometers are used to study inelastic scattering of protons, deuterons, alpha particles, etc. and various transfer reactions, e.g. (d, p), (d, t), (^{3}He, d), (p, t), etc. In former times, the ejectile nuclei were recorded on photographic plates in the focal plane and then scanned visually for exposure tracks as a function of distance along the plate, often presented as 'tracks per 1/4 mm'. Modern devices use position sensitive wire detectors, details of which are beyond the present focus.

Munich was the pre-eminent facility. This is now closed, following many earlier closures of such facilities. From a personal view, advances in nuclear structure studies will slowly come to a stop without such capabilities: there will be an accumulation of leading questions that will simply be unanswerable; questions that used to be answerable, in former times when structural insights were more limited. The Community must not lose sight of a former breadth of view that has now been narrowed, close to a degree that is scientifically unsupportable. In the meantime, we note that elucidation of nuclear structure often depends on old or very old data, which data were not fully appreciated for their potential impact at the time of acquisition. There is grave danger in equating 'old' with: 'bad' or 'irrelevant' or 'superseded'. There is no process in place for in-depth re-evaluation of the entirety of the data pertinent to a single nucleus, let alone a systematic evaluation of an isotopic series, an isotonic series, or a mass region. (The present book is only a minor step towards the need for a much greater effort in this direction. Selected examples of reaction spectroscopy studies of nuclear structure are given in figures 6.14–6.20;

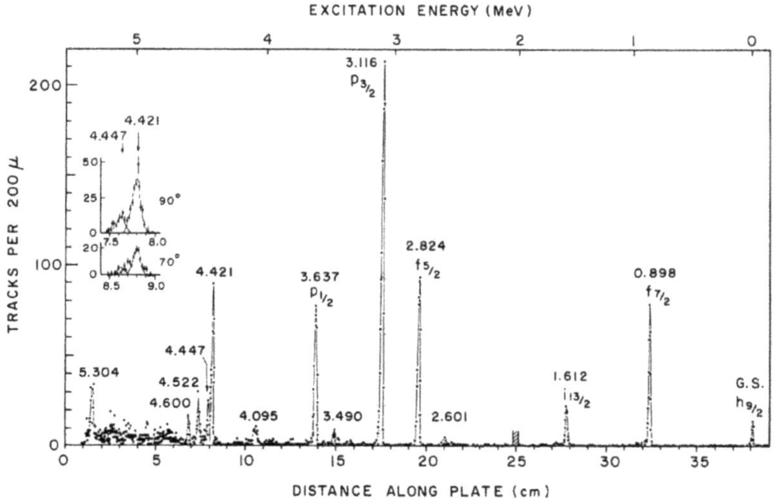

Figure 6.14. Spectrum of deuterons following the $^{208}Pb(^{3}He,d)^{209}Bi$ transfer reaction (cf. figure 4.28). Reprinted with permission from [18]. Copyright (1970) by the American Physical Society.

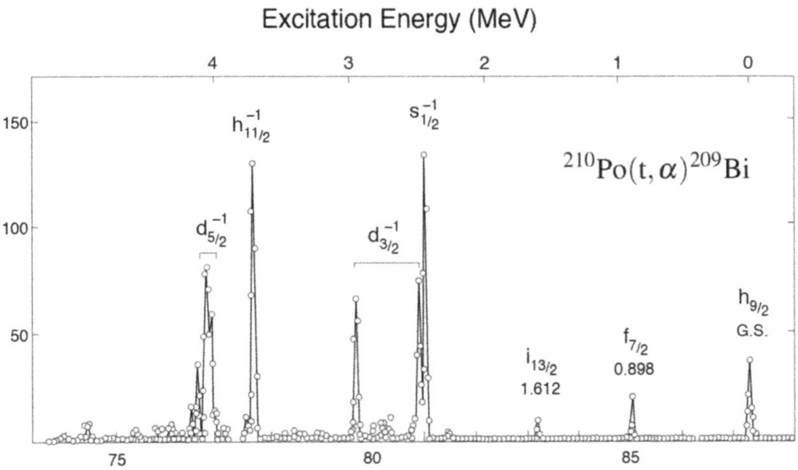

Figure 6.15. Spectrum of ^{4}He ions following the $^{210}Po(t,\alpha)^{209}Bi$ transfer reaction (cf. figures 4.28 and 6.14). Reproduced from [19], copyright (2010) World Scientific Publishing Company. The figure is based on original data found in [20].

note—they are predominantly from work conducted 50 years ago—and yet are the definitive spectroscopic information for detailed structures in these nuclei.

An extension of reaction spectroscopy via detection of ejectile nuclei in coincidence with γ-rays is now quite widely used. This requires detection of the ejectile ion by a semiconductor detector: thin Si wafers are a common choice, usually segmented to obtain some kinematic information for the reaction.

In all of these reaction spectroscopy studies, angular distributions of the ejectile nuclei with respect to the beam direction play a vital role. These distributions have

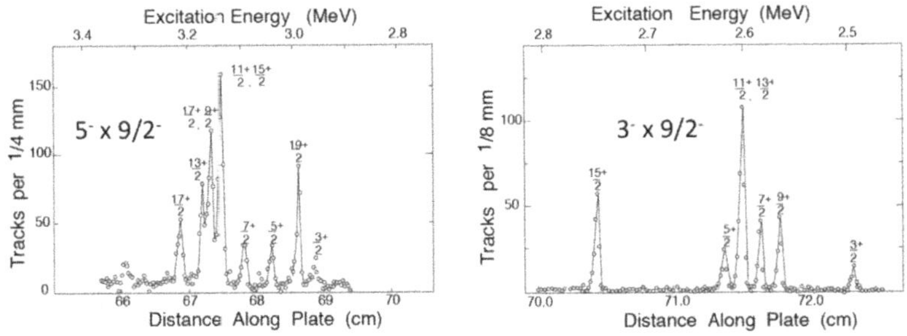

Figure 6.16. Spectrum of deuterons following the ^{209}Bi(d,d')^{209}Bi inelastic scattering reaction (cf. figures 4.28, 4.30, 6.14 and 6.15). The spins, parities, and excitation energies in keV are taken from the Evaluated Nuclear Structure Data File. Reproduced from [19], copyright (2010) World Scientific Publishing Company. The figure is based on original data found in [21].

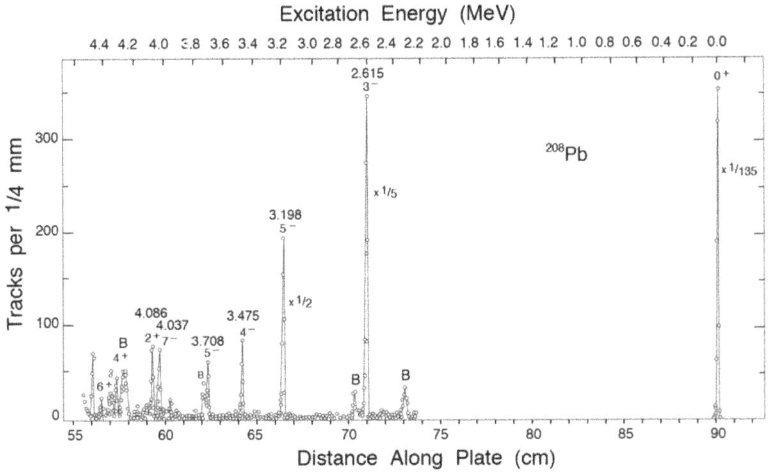

Figure 6.17. Spectrum of deuterons following the ^{208}Pb(d,d')^{208}Pb inelastic scattering reaction (cf. figure 4.28). The peaks labeled 'B' are due to impurities in the target. The spins, parities and excitation energies are from the Evaluated Nuclear Structure Data File. Reproduced from [19], copyright (2010) World Scientific Publishing Company. The figure is based on original data found in [21].

characteristic patterns that permit (limited) angular momentum transfer information: such data lead to spin assignments in the nucleus under study. Even polarized beams are used to resolve ambiguities, e.g. between a pair of shell model states with the same l value but differing j (total spin) values.

Inelastic scattering of electrons stands separately from inelastic scattering of hadronic projectiles, nucleons, nuclei, pions. The reason is that the interaction is purely electromagnetic. Because electron beams have very high velocities for the energies used, the interaction time is very short and so single-step excitation from the ground state dominates the data obtained. Beyond the parallels with Coulomb excitation of light ions, inelastic electron scattering provides so-called 'form factors' for the excitations observed. In simple terms, information on the distribution of

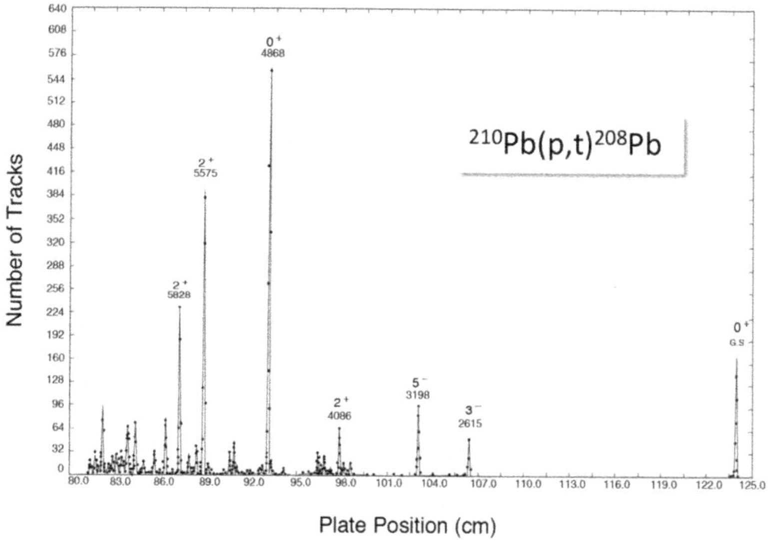

Figure 6.18. Spectrum of tritons following the ^{210}Pb(p,t)^{208}Pb transfer reaction (cf. figures 4.28 and 6.17). Energies below 5 MeV are taken from the Evaluated Nuclear Structure Data File. Reproduced from [19], copyright (2010) World Scientific Publishing Company. The figure is based on original data found in [22].

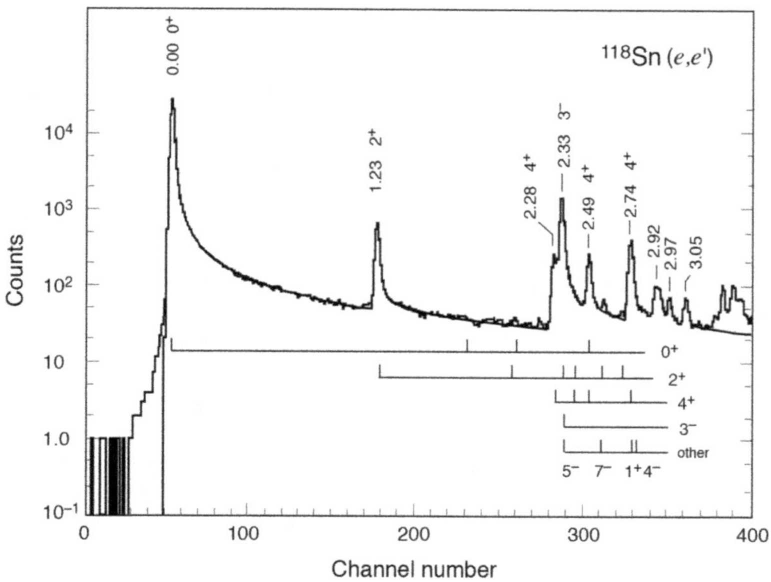

Figure 6.19. Spectrum of electrons following the ^{118}Sn(e,e')^{118}Sn inelastic scattering reaction (cf. figures 4.4, 4.5 and 4.11). Spins, parities, and excitation energies (in MeV) are taken from the Evaluated Nuclear Structure Data File. Reproduced from [19], copyright (2010) World Scientific Publishing Company. The original data appear in [23].

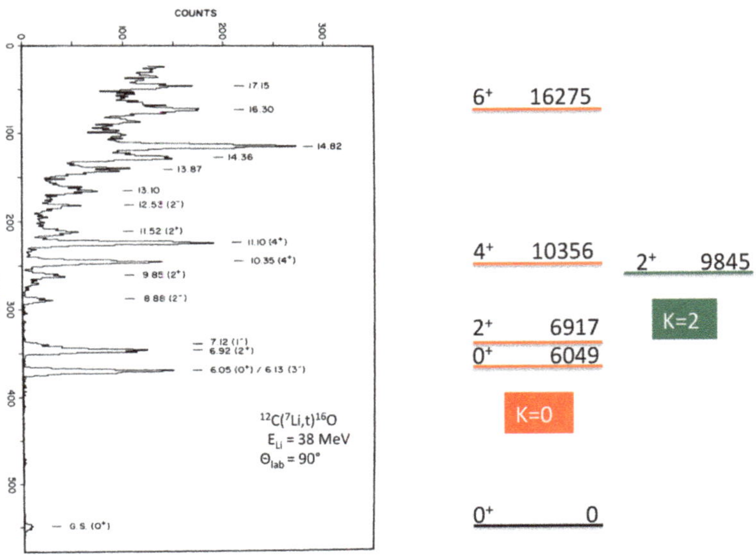

Figure 6.20. Spectrum of tritons following the $^{12}C(^{7}Li,t)^{16}O$ transfer reaction. The 'alpha cluster' addition to ^{12}C supports the interpretation of the band on the 0^{+} state as 4p–4h, π(2p–2h) × ν(2p–2h) configuration (cf. figure 4.27). Reprinted with permission from [24]. Copyright (1976) by the American Physical Society.

contributions to the electromagnetic matrix elements within the nucleus, i.e. as a function of radius, can be obtained.

Finally, we note that inelastic scattering of photons can be used to study nuclei. This is termed nuclear resonance fluorescence. Similar to inelastic electron scattering, it is a purely electromagnetic process and it is confined to single-step excitation from the nuclear ground state. The dominant multipoles are dipole, $E1$ and $M1$. Thus, the technique provides 'maps' of the nuclear dipole responses. An ingenious approach to sources of intense, high-energy, monoenergetic photons is via Compton backscattering of a laser beam by a high-energy monoenergetic electron beam. The HIGS facility at the Triangle Universities Nuclear Laboratory, TUNL provides such capability.

6.8 Atomic hyperfine spectroscopy studies for nuclei

The use of atomic structure to study nuclei is via the so-called hyperfine structure possessed by atoms. Hyperfine structure (structure beyond spin–orbit coupling in the atomic domain) is controlled by nuclear properties—spin, magnetic moment, quadrupole moment, and charge radius (monopole moment).

Atomic hyperfine spectroscopy studies play a major role in characterizing fundamental properties of nuclear states that are sufficiently long lived ($\gtrsim$ 10 ms) that they can be transported into an atomic physics measurement apparatus. In modern times, this means interaction of the isotope, in atomic or ionic form, with a laser beam. The remarkable power of this technique is that, with the high intensity of laser beams, even a single ion can be used repeatedly via multiple interactions to build a high-statistics data set. Further, long-lived ($\gtrsim$10 ms) excited states can be studied in addition to the ground state of selected isotopes.

An aspect of atomic hyperfine spectroscopy that is 'exotic', and is termed 'exotic atom' spectroscopy, is the production of atomic species where the 'orbiting' negative particle is not an electron. This has now been pursued using muons (muonic atom spectroscopy), pions, kaons, sigma baryons, even antiprotons. For the heavier orbiting particles and heavier nuclei the energy regime of the spectroscopy is ~10 keV–5 MeV and is conducted using the techniques of γ-ray spectroscopy. Although the exotic orbiting particle will decay (or annihilate) within a relatively short time, provided it is captured and cascades down through the energy eigenstates of the exotic atom within its lifetime, spectroscopy can be effected. For example, sigma baryons decay with a lifetime of $\sim 10^{-10}$ s which is 'long' compared to the atomic processes. An example of muonic x-ray spectra is shown in figure 6.21.

Exotic atom spectroscopy can be divided into two categories: study of high principal quantum number transitions, where the 'orbiting' particle is 'far away' from the nucleus; and study of low principal quantum number transitions, where so-called isotope effects are large. For example, the first Bohr orbit may be 'inside' the nucleus (recall, for leptons there will not be a Pauli exclusion principle in effect). For mesons and baryons, there will be strong-interaction effects.

6.9 Mass measurements studies for nuclei

The direct measurement of atomic masses has undergone a major revolution with the invention of ion traps, especially Penning traps [27] and storage rings. These studies are now being extended to highly charged ions. (This also results in scaling of hyperfine structure effects by high powers of αZ (where α is the fine-structure constant) and so there is an impact on 'atomic' physics studies of interest.)

By using resonance frequencies of the ion orbits in these various traps and storage rings, extraordinarily high precision on atomic masses has been achieved. The precision in achievable atomic mass is given by

$$\frac{\delta m}{m} \propto \frac{1}{R\sqrt{N}} \tag{6.3}$$

where N is the number of ions involved in the measurement, i.e. the counting statistics, and R is the mass resolving power. The mass resolving power depends on the apparatus employed and, in general, has a strong dependence on the observing time, i.e. for how long the measurement can be conducted—this is naturally limited by the half-life of the isotope of interest (see figure 6.22). This means that long-lived and stable isotopes can be measured with very high precision, e.g. $\delta m/m \sim 10^{-8}$, while short-lived isotopes, which are difficult to produce, can only be measured with precision of order 10^{-6} or even 10^{-5}. While the absolute value of these masses does not warrant the achievable precision, differences in values is the overriding interest. This extends, in the realm of fundamental physics, to the study of differences that should be zero, e.g. the masses of the electron and the positron or the proton and the

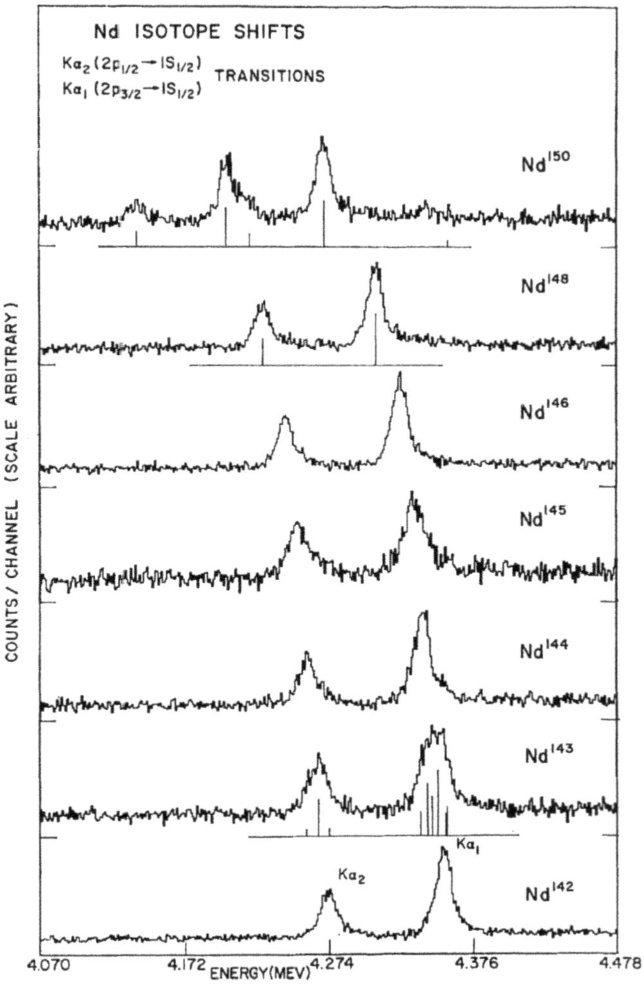

Figure 6.21. Spectrum of photons from muonic atom transitions in ^{142,143,144,145,146,148,150}Nd. These are termed 'muonic x-ray' spectra, but note that they occur in an energy range characteristic of high-energy γ-rays. The isotope shifts are clearly visible (cf. figure 1.4). Note, a muon is 'just inside' the nucleus for the 1s orbital in neodymium ($Z = 60$). For other details, see [25, 26]. Reprinted with permission from [25]. Copyright (1970) by the American Physical Society.

antiproton. Examples of mass measurements, deduced from orbital frequency determinations in a storage ring, are shown in figure 6.23.

6.10 Isotope separation on-line (ISOL) studies for nuclear structure

The choice of including this topic as a dedicated section to this chapter is in recognition of its dominance in modern nuclear structure study[5]. In the early days of nuclear structure study, detailed spectroscopy was limited to study of the

[5] A 'snapshot of the ISOL arena' is presented in [30].

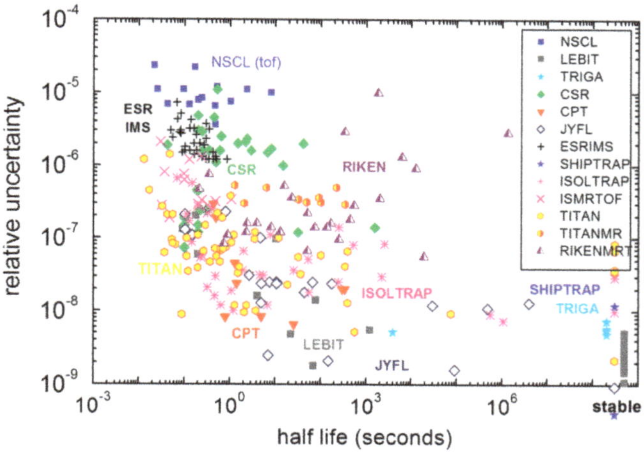

Figure 6.22. Relative uncertainty in atomic mass ($\delta m/m$) measurements as a function of the half-life of the isotope under study. The measurements included in the plot arise from a wide range of different factilities using apparatus such as Penning traps, e.g. LEBIT, CPT, TITAN, ISOLTRAP, SHIPTRAP, JYFL, TRIGA as well as storage rings, e.g. ESRSMS and CSR, in addition to nuclear reaction studies, e.g. SPEG. Reproduced from [28] with permission of Springer-Nature, copyright (2019).

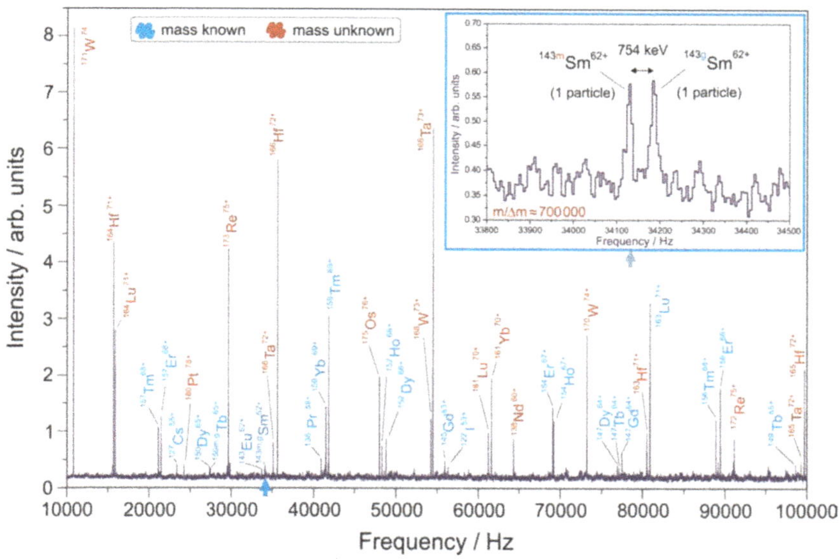

Figure 6.23. Orbital frequency spectra of single ions in an ion-storage ring. The ions are highly stripped and the orbit frequency is determined by charge/mass ratios. The inset shows the difference between two states in $^{143}Sm^{62+}$ ions that, because they differ in mass-energy by 754 keV, exhibit slightly different orbit times. The ring is 108 m in circumference; the signals are acquired by multiple passes of a single ion of a given charge/mass. Here, the ^{143}Sm ions are passing with a frequency of ~34 150 Hz. The method is termed Schottky mass spectrometry and this refers to the passage of ions imparting an electrical response in conducting plates. Ions are generally 'lost' from the beam when their charge state changes due to pickup of an electron from a stray gas molecule in the vacuum. Reprinted from [29], copyright (2004) with the permission of Elsevier.

stable isotopes and isotopes immediately adjacent to them. While radioactive decay scheme study could provide a limited view of nuclear structure, this was not a detailed view. With in-beam spectroscopy, our structure horizons expanded, but with the limitations that have already been noted herein.

A major initiative to expand our nuclear structure horizons was launched in 1967 at a workshop in Lyseskil in Sweden. This led to the ISOLDE Project: the use of an on-line isotope separator for the provision of intense beams of radioactive isotopes. There have been many emulations of the basic ISOLDE idea, and a major extension to fragment separators. The key to the ISOL method is the use of ion sources: often this has been by trial-and-error with respect to delivery of a particular elemental species[6], to the degree that the term 'ion-sourcery' has been coined, with no intention of humor.

Generally, radioactive species are extracted as singly charged ions from an ion source, where the recoil of the species into the ion source follows a nuclear reaction. The extracted species is then mass separated magnetically and transported, either as an ion using an ion-optic system or as an atom on a moving tape system, even as an atom or ion in a supersonic gas jet prior to mass separation and secondary transport, to a measurement station. Construction of such transport systems has been a major technological challenge in itself (and still continues in a developmental mode): as a universal constraint, high tape speeds are desirable, *with no stretching of the tape*[7], for continuous running times of days.

Ion-source development has undergone its own set of developmental steps, from thermal ionization, through electron bombardment and so-called cyclotron resonance ionization, to laser induced ionization. Remarkably, specific isomeric species of an isotope can now be separated via its atomic hyperfine structure and selective ionization using lasers.

The power of the ISOL philosophy lies in the use of the radioactive beam as the species of interest (radioactive targets have an obvious limitation, far from stability the short half-lives would lead to unacceptably high specific activities and a limit tied to production rate and target decay). The Coulomb excitation of radioactive beams is now a realized prospect for major advances in nuclear structure study.

We make a word of caution to the Reader who does not work in this arena. Production of isotopes far away from the line of beta stability decreases very dramatically—typically an order of magnitude per step of one unit of Z in an isobaric chain. Thus, each new facility generation usually only extends members of an isobaric chain by one unit in Z. The data obtained are sparse and the spectroscopic detail is extremely limited. Frequently, data for a 'new' isotope are limited to a mass, a charge radius, a half-life, and a first excited state. Often the first excited state is only characterized by its energy, i.e. lacking a direct spin and parity determination. Then models and systematics take over. While this is a 'safe' way to proceed when structures are dominated by one type of 'coupling scheme'; when

[6] We especially note the major pioneering work of Helge Ravn, ISOLDE-CERN and Reinhardt Kirchner, GSI-Darmstdt, in these endeavors.

[7] If the tape stretches, eventually the sources do not stop in front of the detectors.

sudden changes occur, e.g. deformed structures intrude to become the ground state in a region where spherical shapes are the norm, elucidation of the structural details can take ~25 years until a consensus view is reached.

6.11 Lifetime measurements by Doppler techniques

The Doppler effect on a γ-ray energy can be used to deduce the lifetime of the nuclear state from which the γ-ray is emitted. Always, the nucleus of interest is produced or extracted in a reaction and is moving at a high velocity after this has occurred. This method has a key proviso: one must characterize the feeding history of the level in question. A simple situation is depicted in figure 6.5(e), (f): the levels are populated with a (neutron) beam that has energy just above threshold for the level population. Thus, the feeding history is 'prompt'. Delayed feeding, whether via longer-lived levels or shorter-lived levels requires careful de-convolution of the feeding history.

Detection of the Doppler effect can be direct, as in figure 6.5(e), (f) and as also depicted in figure 3.13, or it can be less direct as in the so-called 'recoil distance Doppler-shift method', RDDS method or RDM. The RDM involves a so-called 'plunger' that plays the role of stopping the recoiling nucleus of interest: Doppler energy shifts are mapped by precisely controlled (micrometer) shifts in the position of the plunger relative to the target (production point of the species of interest). Feeding history can be mapped by coincidence gating of the γ-ray under study by selection of other γ-rays in the decay scheme (which must first be elucidated using a separate dedicated study). A sample set of RDM data are shown in figure 6.24. Further details will be covered later in the Series.

6.12 Lifetimes of states associated with radiative emission of γ-rays

The radiative emission of γ-rays from excited states in nuclei is a spontaneous process. It results from the coupling of the nuclear electric charge to the electromagnetic vacuum. Some details of this were noted in chapter 2 with respect to the so-called Weisskopf estimates for radiative strength. Equations (6.4) and (6.5) give general relationships for electric and magnetic transitions for all low multipolarities, $\lambda = 1, 2, 3, \ldots$.

$$B(E\lambda; i \to f) = \frac{\mathcal{F}_{\lambda}^{(E)}}{(E_{\gamma})^{2\lambda+1} T_{1/2}^{\gamma}(E\lambda) 1.442\,7} e^2 b^{\lambda} \tag{6.4}$$

$$B(M\lambda; i \to f) = \frac{\mathcal{F}_{\lambda}^{(M)}}{(E_{\gamma})^{2\lambda+1} T_{1/2}^{\gamma}(M\lambda) 1.442\,7} \mu_N^2 b^{\lambda-1}, \tag{6.5}$$

The constants are given in table 6.1; E_{γ} values are in units of MeV, $T_{1/2}$ values are in units of picoseconds, ps, e is the unit of electric charge, b is the unit of area in barns ($= 10^{-24}$ cm^2) and μ_N is the nuclear magneton. The multipolarity is a function of the angular momentum change due to photon emission. Besides an intrinsic spin of $1\hbar$,

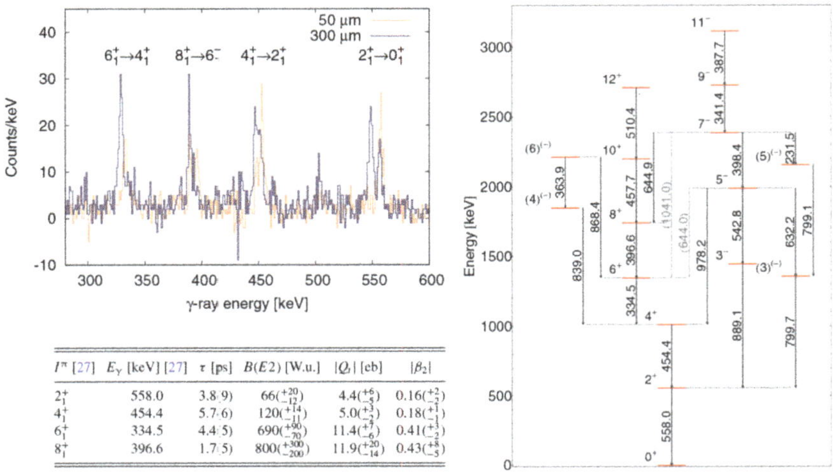

I^{π} [27]	E_{γ} [keV] [27]	τ [ps]	$B(E2)$ [W.u.]	$\vert Q_t \vert$ [eb]	$\vert\beta_2\vert$
2_1^+	558.0	3.8(9)	$66(^{+20}_{-12})$	$4.4(^{+6}_{-5})$	$0.16(^{+2}_{-2})$
4_1^+	454.4	5.7(6)	$120(^{+14}_{-11})$	$5.0(^{+3}_{-2})$	$0.18(^{+1}_{-1})$
6_1^+	334.5	4.4(5)	$690(^{+90}_{-70})$	$11.4(^{+7}_{-6})$	$0.41(^{+3}_{-2})$
8_1^+	396.6	1.7(5)	$800(^{+300}_{-200})$	$11.9(^{+20}_{-14})$	$0.43(^{+8}_{-5})$

Figure 6.24. Doppler line shifts for γ-rays observed in the ^{103}Rh(^{78}Kr, p2n)^{178}Hg reaction. The γ-ray spectra are for the largest and smallest plunger distances. The tabulation is for the deduced lifetimes of the transitions for which data were obtained: deduced $B(E2)$ values, and quadrupole moments and quadrupole deformation parameters inferred from these $B(E2)$ values, are shown. The ^{178}Hg level scheme shows the locations of the transitions. (Q_t designates that it is a 'transition' quadrupole moment, i.e. as one would deduce using equations (3.6) and (3.8), whence $Q_t = Q_0$.) Reprinted with permission from [31]. Copyright (2019) by the American Physical Society.

Table 6.1. The constants in equations (6.4) and (6.5), particular to electric and to magnetic radiative decays. These constants are dimensionless.

λ	1	2	3	4	5
$\mathcal{F}_{\lambda}^{(E)}$	6.29×10^{-6}	8.162×10^{-2}	1.751×10^{3}	5.896×10^{7}	2.894×10^{12}
$\mathcal{F}_{\lambda}^{(M)}$	5.689×10^{-2}	7.383×10^{2}	1.586×10^{7}	5.332×10^{11}	2.617×10^{16}

photons can change spin-angular momentum by integer amounts; with the proviso that spin-angular momentum is a vector quantity.

Table 6.2 gives some examples of γ-ray transitions of various multipolarity, with associated half-lives for the decay process. The extreme variation in lifetimes should be noted. This variation reflects the fact that photon emission is not an effective mechanism for 'removing' angular momentum from a nucleus. Table 6.2 also gives the total internal-conversion coefficient, ICC for each transition. The value of an ICC depends sensitively on Z, E_{γ} and multipolarity; it does not depend on the Weisskopf strength of the transition. Note that the total half-life observed for a level is the result of both γ-ray emission and internal conversion (summed over all atomic shells): if internal conversion did not occur, the half lives would be longer; cf. equations (2.5) and (2.6).

Table 6.2. A selection of electromagnetic transition strengths (in Weisskopf units, W.u.), which illustrates multipolarities $\sigma = E$ or M, $\lambda = 1, 2, 3, 4$. Transition strengths marked with † are amongst the strongest known transition strengths for that multipolarity. Where internal conversion significantly modifies the half-life, the total conversion coefficient, α_{total} is given. Multiple examples are shown, where of illustrative interest. The data are taken from ENSDF. Note that from data such as Coulomb excitation and inelastic electron scattering, there are localized mass regions with significant $E3$ [32] and $E4$ [33] strength.

Isotope	Transition energy	Multipolarity	$T_{1/2}$	α_{total}	$B(\sigma\lambda)$
	(keV)				W.u.
^{11}Be	320.0	E1	115 fs		3.6×10^{-1} †
^{16}O	7 116.9	E1	8.3 fs		3.5×10^{-4}
^{16}O	6 917.1	E2	4.70 fs		3.1
^{208}Pb	4 085.5	E2	0.80 fs		8.4
^{24}Mg	1 368.7	E2	1.33 ps		21.5
^{168}Er	79.8	E2	1.853 ns	7.04	213
^{250}Cf	42.7	E2	96 ps	1293	340 †
^{16}O	6 129.9	E3	18.4 ps		13.5
^{208}Pb	2 614.5	E3	16.7 ps		33.8 †
^{52}Fe	597.1	E4	45.9 s		4.6×10^{-4}
^{117}Sn	158.6	M1	279 ps	0.155 8	1.7×10^{-2}
^{43}Sc	151.8	M2	439 μs	0.040 6	6.91×10^{-2}
^{113}Sn	77.4	M3	21.4 min	181	2.8×10^{-2}
^{119}Sn	65.7	M4	293.1 d	5000	5.2

6.13 Exercises

The relative challenge and time investment associated with the exercises is indicated from * to ***. Exercises marked with '**E**' require interactive access to figures or watching included video exercises and will only have such functionality in the e-book version.

6-1. The Bohr radius is that for an electron in its lowest state in a Hydrogen atom and is given by

$$a_0 = \frac{4\pi\epsilon_0\hbar^2}{m_e e^2} \tag{6.6}$$

where m_e is the electron mass. The corresponding radius for the first Bohr orbit for an element with higher Z is given by

$$r_Z = \frac{a_0}{Z}. \tag{6.7}$$

(a) What would be the radius of the first Bohr orbit for a muon bound to ^{150}Nd ($Z = 60$)? **

(b) Is this orbit, on the average, 'inside' of the nucleus? **

(c) Using the concept of the first and second Bohr orbits, can the energy scale in figure 6.21 be simply explained? **

6-2. The energy needed to dissociate the HCl molecule is 4.62 eV. Considering the precision of the masses of the H atom and a ^{37}Cl atom given in the https://www-nds.iaea.org/amdc/ AME, can it be said that 'we can weigh the mass-energy of the chemical bond'? **

6-3.

(a) Using table 6.1, obtain the scale reduction factors for $E1 \rightarrow E2 \rightarrow E3 \rightarrow \cdots$and compare to $M1 \rightarrow M2 \rightarrow M3 \cdots$. *

(b) Estimate the $\mathcal{F}$ factor for an $M8$ transition. *

(c) Estimate the half-life for decay of the 1147 keV state in ^{178}Hf (cf. figure 3.21) direct to the ground state for a transition strength of 1 W.u. **

6-4. The precision on nuclear masses which can be obtained depends on the number of ions in the measurement, N, and the mass resolving power, R, and is given by equation (6.3). The mass resolving power, R, for a Penning trap is typically 10^5 for an observation time of 100 ms. (This means that the isotope under study must be sufficiently long-lived to allow a measurement to be conducted for 100 ms.) Estimate the number of ions that would need to be trapped to achieve a mass precision, $\delta m/m = 10^{-7}$. *

6-5. The isotope ^{176}Lu occurs naturally. Explain this in terms of expectations based on masses of $A = 176$ isobars and spins of states in ^{176}Lu and its neighbors. ***

6-6. The isotopes ^{235}U and ^{238}U both occur naturally. They possess relative natural abundances of 0.720 4% and 99.274 2% and half-lives of 7.04×10^8 y and 4.468×10^9 y, respectively. Assuming that these two isotopes of uranium were present in equal amounts when they were formed (in some coalescing binary neutron-star pair, far-far away, long-long ago), estimate how long ago that was. (Compare with the estimated age of the Earth: 4.5×10^9 y.) *

6-7. The longest-lived isotope of plutonium is ^{244}Pu with a half-life of 8.13×10^7 y. Assuming that plutonium and uranium have similar mineral chemistry and were formed in equal amounts in the scenario posed in exercise 6-6, estimate the ratio for ^{244}Pu/^{238}U in uranium ore. Has this been investigated? *

6-8. Explore other 'cosmochronometer' candidates, and sensitivity to trace isotope detection using mass analyses. ***

6-9. There is only one reported example of an $E6$ transition, in ^{53}Fe. Assess the credibility of the data on which this assignment is made (use the https://www.nndc.bnl.gov/ensdf/ ENSDF compilation and evaluation). ***

References

[1] Alder K, Bohr A, Huus T, Mottelson B and Winther A 1956 Study of nuclear structure by electromagnetic excitation with accelerated ions *Rev. Mod. Phys.* **28** 432

[2] Allmond J M *et al.* 2015 Investigation into the semimagic nature of the tin isotopes through electromagnetic moments *Phys. Rev.* C **92** 041303

[3] Hardy J C, Carraz L C, Jonson B and Hansen P G 1977 The essential decay of pandemonium: A demonstration of errors in complex beta-decay schemes *Phys. Lett.* B **71** 307
[4] Morinaga H and Gugelot P C 1963 Gamma rays following (α,xn) reactions *Nucl. Phys.* **46** 210
[5] Bertram Hansen G, Elbek B, Hagemann K A and Hornyak W F 1963 Population of rotational states by proton induced reactions *Nucl. Phys.* **47** 529
[6] Stephens F S, Lark N L and Diamond R M 1965 Rotational states produced in heavy-ion nuclear reactions *Nucl. Phys.* **63** 82
[7] Twin P J *et al.* 1986 Observation of a discrete-line superdeformed band up to 60 $\hbar$ in ^{152}Dy *Phys. Rev. Lett.* **57** 811
[8] Broda R, Wrzesinski J, Michelagnoli C, Lunardi S, Ur C, Bazzacco D, Menegazzo R, Mengoni D and Recchia F 2020 Shower of γ rays in the decay of the 49/2$^+$ isomeric state in ^{147}Gd *Phys. Rev.* C **101** 064320
[9] Venhart M *et al.* 2020 Population of a low-spin positive-parity band from high-spin intruder states in ^{177}Au: The two-state mixing effect *Phys. Lett.* B **806** 135488
[10] Kulp W D *et al.* 2008 Search for intrinsic collective excitations in ^{152}Sm *Phys. Rev.* C **77** 061301
[11] Kulp W D *et al.* 2007 $N = 90$ region: The decays of ^{152}Eu^{m,g} to ^{152}Sm *Phys. Rev.* C **76** 034319
[12] Zamfir N V *et al.* 1999 $B(E2)$ values and phase coexistence in ^{152}Sm *Phys. Rev.* C **60** 054312
[13] Schmelzenbach P *et al.* 2018 γ-ray spectorscopy of ^{150}Sm through the β decay of ^{150}Pm (t = 2.7 h) and ^{150}Eu^{m} (t = 12.8 h) *Phys. Rev.* C **98** 034311
[14] Venhart M *et al.* 2017 Application of the Broad Energy Germanium detector: A technique for elucidating β-decay schemes which involve daughter nuclei with very low excited states *Nucl. Instrum. Methods Phys. Res.* A **849** 112
[15] Kibédi T *et al.* 2020 Radiative width of the Hoyle state from γ-ray spectroscopy *Phys. Rev. Lett.* **125** 182701
[16] Rissanen J, Elomaa V V, Eronen T, Hakala J, Jokinen A, Rahaman S, Rinta-Antila S and Äystö J 2007 Conversion electron spectroscopy of isobarically purified trapped radioactive ions *Eur. Phys. J.* A **34** 113
[17] Roussière B *et al.* 2000 High-resolution measurements of low-energy conversion electrons *Hyperfine Interact.* **129** 119
[18] Ellegaard C, Patnaik B and Barnes P D 1970 Fragmentation of the single-particle strength in ^{209}Bi *Phys. Rev.* C **2** 2450
[19] Rowe D J and Wood J L 2010 *Fundamentals of Nuclear Models: Foundational Models* (Singapore: World Scientific)
[20] Barnes P D, Romberg E, Ellegaard C, Casten R F, Hansen O, Mulligan T J, Broglia R A and Liotta R 1972 Proton-hole states in ^{209}Bi from the ^{210}Po(t,α) reaction *Nucl. Phys.* A **195** 146
[21] Ungrin J, Diamond R M and Tjø P O 1971 Inelastic deuteron scattering in the lead region *Mat. Fys. Medd. Dan. Vid. Selsk.* **38** 34
[22] Igo G J, Barnes P D and Flynn E R 1970 Mixing of two-particle, two-hole states in ^{208}Pb *Phys. Rev. Lett.* **24** 470
[23] Peterson R J, Kraushaar J J, Braunstein M R and Mitchell J H 1991 Inelastic electron scattering to collective states of ^{118}Sn *Phys. Rev.* C **44** 136
[24] Cobern M E, Pisano D J and Parker P D 1976 Alpha-transfer reactions in light nuclei. III. (^{7}Li,t) stripping reaction *Phys. Rev.* C **14** 491
[25] Macagno E R *et al.* 1970 Muonic Atoms. II. Isotope shifts *Phys. Rev.* C **1** 1202

[26] Hitlin D, Bernow S, Devons S, Duerdoth I, Kast J W, Macagno E R, Rainwater J, Wu C S and Barrett R C 1970 Muonic Atoms. I. Dynamic hyperfine structure in the spectra of deformed nuclei *Phys. Rev.* C **1** 1184

[27] Dilling J, Blaum K, Brodeur M and Eliseev S 2018 Penning-trap mass measurements in atomic and nuclear physics *Ann. Rev. Nucl. Part. Sci.* **68** 45

[28] Lunney D 2019 New mass measurements with trapped (radioactive) ions and related fundamental physics *Hyperfine Interact.* **240** 48

[29] Münzenberg G 2004 Storage-ring experiments with exotic nuclei: from mass measurements to the future *Prog. Part. Nucl. Phys.* **53** 351

[30] Jenkins D G 2014 Recent advances in nuclear physics through on-line isotope separation *Nat. Phys.* **10** 909–13

[31] Müller-Gatermann C *et al.* 2019 Shape coexistence in ^{178}Hg *Phys. Rev.* C **99** 054325

[32] Kibédi T and Spear R H 2002 Reduced electric-octupole transition probabilities, $B(E3;\, 0_1^+ \rightarrow 3_1^-)$–an update *At. Data Nucl. Data tables* **80** 35

[33] Sheline R K *et al.* 1999 Systematic investigation of hexadecapole collectivity in even–even nuclei *Czech. J. Phys.* **49** 1047

IOP Publishing

Nuclear Data

A primer

David G Jenkins and John L Wood

Chapter 7

Metrology for nuclear structure study

The study of nuclear structure depends on nuclear spectroscopy, which involves the measurement of energies and intensities of radiations emitted by nuclei and of particle beams following nuclear reactions. There are observations of the angular and time dependent distributions of radiations from nuclei in magnetic fields, which yield magnetic moments. There are observations of intensity as a function of time to establish half-life values, which lead to magnitudes of transition matrix elements. There are observations of Doppler broadened line shapes from which life times can be deduced, extending transition matrix element information to very short-lived states. Nuclear structure study also has a dependence on atomic spectroscopy by way of optical hyperfine spectroscopy. Even, exotic atoms such as muonic or pionic atoms can be produced and their properties studied to yield nuclear structure information. Further, ions can be trapped and orbital resonances can be studied to yield values of masses. All of these issues necessitate a familiarity with metrology, in the form of calibration standards, units, and definitions of measurable quantities.

Energy is in principle straightforward to standardize. The Joule is easy to conceptualize: literally it can be directly experienced as heat, and this can be related to work. James Joule's quantification between work and heat is a time-honored historical step in the history of physics. However, γ-ray energies are usually determined by collecting electrical charge produced by ionization in a detector, and measured as a voltage. Occasionally, γ-ray energies are determined by diffraction, and then detecting them at a precisely defined diffraction angle (which equates to the energy of the radiation). While 'heat' seems very remote from γ-ray 'energy' measurements, bolometers are employed for certain ultra-precise work at very low energies in radionuclide study.

Intensity is a more subtle quantity, while it can be directly experienced as visible light, quantification depends on detector 'efficiency'. Efficiency depends on quantum mechanisms of interaction such as the photoelectric effect; for the detection of γ-rays, interactions also extend to Compton scattering and electron–positron pair formation.

doi:10.1088/978-0-7503-2674-2ch7

Herein, we sketch some key steps in the chain from basic metrology to practical standards for nuclear spectroscopy. The framework of metrology—from the Système International (SI) Base Units to units, from fundamental constants to the processes of physics—is an intricate and historically significant framework of 'connections'.

Energy is normally quantified in Joules. In nuclear spectroscopy we use multiples of the electronvolt: the kinetic energy change of an electron (one unit of electrical charge) when it undergoes a change of one volt in its potential energy, e.g. moving in an electrical field gradient. Thus, we depend on the metrologically defined equivalence between the Joule and the electronvolt, 1 eV $=1.602\ 176\ 634 \times 10^{-19}$ J (and therefore, on the definition of the volt); albeit, in practice we work entirely within the definition of the electron volt. Electron volts apply even when we speak of nuclear masses [where the unified atomic mass unit, also called the Dalton, is quoted as 931.494 102 42(28) MeV].

Intensity is quantified in candelas. In nuclear spectroscopy we can circumvent issues of source intensity because we can control the number of nuclei in the source. Thus, we measure the decay branches from specific initial states, e.g. the radioactive decay of a nuclear ground state to various states in a daughter nucleus, and from knowledge of the half-life, arrive at an effectively calibrated radiation intensity source. Thus, we depend almost entirely on radioactive decay standards for efficiency calibrations to determine radiation intensities. However, we note that so-called Monte Carlo simulations, for the interaction of radiations with active volumes of detectors, now fully match in precision the determination of detector efficiencies using radioactive decay standards. Disintegration rates of radioactive sources are measured in Becquerels or Curies.

The attractiveness of radioactive decay standards is that they handle both energy and efficiency calibrations simultaneously. Their limitation is that they only offer a limited set of calibration points. This devolves onto a serious limitation for efficiency calibrations: the inter-calibration of multiple radioactive source intensities. Or for both energy and intensity calibration: there is a lack of sources that offer high-energy calibration lines.

The highest available precision for γ-ray energies is via measurement of their wavelengths as manifested in the GAMS4 program at Institut Laue-Langevin (ILL) in Grenoble, France. (See [1]; and see, e.g. [2] and [3].)

The highest available precision for γ-ray intensities is in association with quantification of so-called Fermi β-decay branching ratios [4, 5].

Universally, nuclear spectroscopic information is stored in digital form and so the data acquired in experiments must be amenable to conversion from 'analog' to 'digital'.

We give brief introductions to the SI base units, derived units of importance to nuclear structure study, and fundamental constants of relevance to nuclear structure study.

7.1 The SI base units

The SI Base Units are a remarkable distillation of science and technology into seven defined quantities. There has been an historical evolution, adjudicated by the Consultative Committee for Units (CCU), which has moved this set of quantities away from artifacts, to definitions in terms of Nature's fundamental constants. This process towards definitions in terms of fundamental quantities of Nature was completed in 2019. They are portrayed in figure 7.1.

Historically, the base units of physics can be said to have started with Isaac Newton and his formulation of mechanics and gravity; thus, leading to concepts of mass, length, time, and a constant to describe the law of gravity. However, to communicate in quantitative terms, units needed to be agreed upon. It was the French Revolution that gave us the foundation of our system of International Units—the Metric System. (Although, some of us manage to get by in every day life with chains, stones, fortnights, and even some other units, cf. figure 7.2.) This changed with the emergence of electric and magnetic phenomena. It was Gauss who in 1832 set Society on the course to an appreciation and the need to think more deeply about the interconnectedness of measurable quantities when confronted with electromagnetism. But it took until 1960 to formally adopt the m–kg–s–A system, which became the SI. With the addition of temperature, brightness of light, and counting of atoms, the modern SI has been reached. However, that is not exactly the way we now define the Base SI Units, as the following will reveal.

The saga by which we have arrived at the modern SI has been told in many places, so we only touch on a few key concepts. In earliest times, the meter and the kilogram were artifacts and the second was 1/86 400 of a day. Evolution of clocks brought the

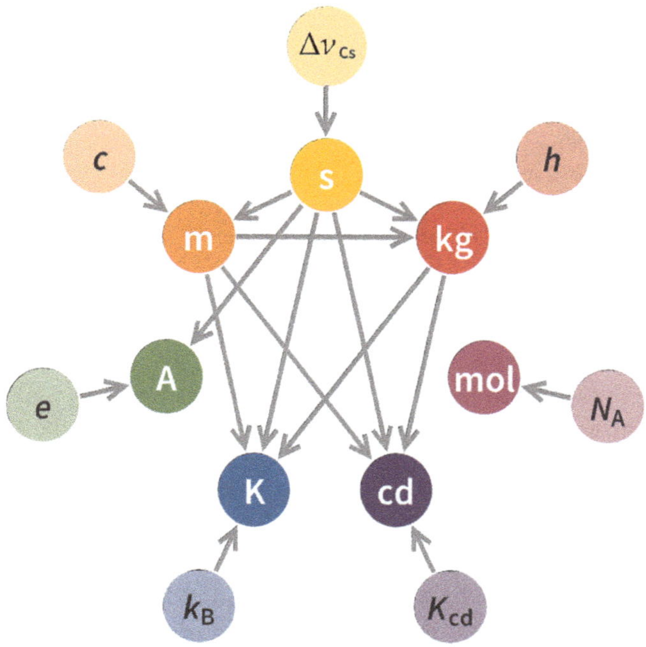

Figure 7.1. The 2019 redefinition of the SI units and their connection to fundamental constants. This image has been obtained by the authors from the Wikimedia website where it was made available under a CC BY-SA 4.0 licence. It is included within this chapter on that basis. It is attributed to Emilio Pisanty.

Figure 7.2. Units of length used in England, ca. 1809. The conversion factors are 1 M = 8 F, 1 F = 40 P. The mile (M), the furlong (F), and the pole (P, also known as the rod or perch or lug) all have ancient origins: mile—mille (Latin, one thousand paces), furlong—furh lang (Anglo-Saxon or Old English, furrow length), perch—pertica (Latin, via perche from French, pole or staff). The subunits have their origin in farming where a primary interest is area, e.g. the acre (Latin—ager, field) is defined by a rectangle 1 F × 4 P (we note that 4 P = 1 C, a chain). For the interested Reader, we invite them to explore the origin of knot, hundredweight, and nanowales. We note, nuclear physicists concern themselves with area: barns.

atomic clock to the world of metrology; first envisaged by Isidore Rabi in 1945, and conceptually realized as the ammonia maser in 1949. It was the caesium atomic clock, the first of which was built by Louis Essen in 1955 at the National Physical Laboratory, NPL in the UK, that was used in 1960 to define the modern second as exactly 9 192 631 770 oscillations corresponding to the ground-state hyperfine splitting frequency of the caesium atom.

Remarkably, and well known to many, the speed of light in a vacuum is the same for all observers. The length artifact was replaced for a while by an optical wavelength (in the ^{86}Kr atom); but with the cesium clock this was replaced by one meter being the distance travelled by light in 1/299 792 458 s, exactly. This was recognized in 1975, adopted in 1983, and remains with us today. However, the precision of atom-based clocks has advanced far beyond the capabilities of those based on oscillations in caesium atoms, and we can anticipate that some day we will not be talking of oscillations in caesium atoms.

The real drama arrived with the replacement of the International Prototype of the Kilogram (IPK), 'Le Grand K', made official on May 20, 2019[1]. This replacement is a new conceptual role for Planck's constant, and this needs some explanation. At the heart of this epoch-defining step is the so-called Kibble balance[2], which can observe a mass (say 1 kg) under the influence of local gravity and counteracted by controlled electromagnetic forces.

A Kibble balance operates in two modes. In the first mode, a mass is balanced by the force produced by an electrical current flowing in a coil surrounded by a magnetic field. Equating the Lorentz force of the magnetic field on the coil to the local attraction of gravity on the mass yields the relationship

$$\text{weight} = mg = BLI, \tag{7.1}$$

where m is the mass in kilograms, g is the local acceleration due to gravity, B is the magnetic field strength, L is the length of the wire in the coil, and I is the current in amps flowing through the coil. In the second mode, the same coil is moved through the magnetic field at a known speed: by Faraday's law of induction, a potential is induced between the ends of the wire in the coil, given by

$$U = BL\mathrm{v}, \tag{7.2}$$

where U is the potential difference in volts, B and L are the same as in equation (7.1), and v is the speed in m s^{-1}. The common factor BL is eliminated, yielding

$$mg\mathrm{v} = UI. \tag{7.3}$$

Both sides of equation (7.3) have dimensions of power, and the right-hand side, volts × amps = watts, imparted the original name to such a device, i.e. the watt balance.

The connection to Planck's constant in the above relationships comes about through two extraordinary macroscopic quantum mechanical effects. The first is the so-called 'AC Josephson effect', which relates a voltage to a frequency across a tunneling gap between two superconductors via

$$h\nu = 2eV, \tag{7.4}$$

where h is Planck's constant, ν is a frequency (of electromagnetic radiation, emitted or absorbed), e is the fundamental unit of electrical charge, and V is measured in volts. For $V = 1$ volt, $\nu = 483,\ 597.848\,4 \cdots$GHz, i.e. em radiation of wavelength 62 μm, which is in the microwave region of the em spectrum. The second is the so-called 'integer quantum Hall effect', which relates the appearance of a transverse potential difference, induced by a longitudinal potential difference and current,

[1] An account of the details can be found in [6].

[2] The Kibble balance was envisaged in 1975 by Bryan Kibble at the NPL as a means to precision measurement of magnetic moments. It evolved into the so-called watt balance, as described, e.g. in [7].

acting in a two-dimensional semiconductor of finite dimensions under the influence of a magnetic field perpendicular to the plane of the semiconductor. This translates into a transverse resistance, which connects the transverse potential difference to the longitudinal current (which in turn is defined by the longitudinal potential difference and the longitudinal resistance), and is given by

$$R_H = \frac{h}{ie^2}, \tag{7.5}$$

where R_H is the transverse or quantized Hall effect resistance, h is Planck's constant, e is the fundamental unit of electrical charge, and $i = 1, 2, 3, \ldots$, i.e. R_H is quantized in units of h/e^2 which equates to $25, 812.80745\cdots$ ohms. The quantization arises from the number of completely filled Landau levels[3] in the semiconductor device. This effect was first observed in 1980, by Klaus von Klitzing. Using standard relationships between potential difference, resistance, and current, with reference to multiples of the Josephson voltage and multiples of the von Klitzing resistance, the relationship from the Kibble balance provides a definition of the kilogram in terms of Planck's constant. The adopted value of Planck's constant is $h = 6.626\ 070\ 15 \times 10^{-34}$ J s (exactly).

The ampere is related to the unit of electrical charge via 1 amp.second = 1 As = 1 Coulomb (C) and 1 $e = 1.602\ 176\ 634 \times 10^{-19}$ C (exactly). The Kelvin, K is defined in terms of the Boltzmann constant, k as $k = 1.380\ 649 \times 10^{-23}$ J K^{-1} (exactly). In the sub-field of nuclear physics, as already noted, we do not use the candela unit for radiation intensity.

The seven base units of the SI are independent of each other, i.e. a change in the numerical value any one of them will not change the numerical value of any of the others.

The foregoing leads us to the values of the fundamental physical constants, and so we follow with some details of particular interest to nuclear physicists.

7.2 The fundamental physical constants

Physical quantities that are constant span a very large catalog. Most prominent in this catalog there are quantities such as the speed of light (in vacuo), Planck's constant, the unit of electrical charge, and the Newtonian gravitational constant, which are familiar to all who have been 'introduced' to physics in its modern form. A few steps further into physics and one learns that the masses of all particles both fundamental and composite are also constants. Thus, in particular, the mass of the electron, the proton and the neutron are encountered as among leading manifestations of mass that are universal in Nature. When dealing with the fundamental manifestation of forces, electromagnetism must be quantified with the permittivity and permeability of the vacuum. When relating the atomic scale of matter to its macroscopic manifestations, the Avogadro number and the Boltzmann constant are encountered. The specialist learns that there are strong (or color) and weak charges characterizing the interactions with those names. Even, that Nature has chosen to

[3] The quantum mechanics of Landau levels is developed in full detail in [8], chapter 13 of the Series.

'flavor' its basic building blocks, but that both quarks and neutrinos are manifested in flavor-mixed form, characterized by a set of constants; and yet further Nature has found a way for this flavor mixing to violate fundamental symmetries of space, time, and matter–antimatter in certain combinations. At the end of this, on-going saga, we find that there is a Standard Model with 26 parameters, and it is anticipated that more will be needed when we 'understand' Dark Matter, Dark Energy, and maybe new levels of substructure in our Universe. Herein, we aim to isolate those measures of this fabric of physics that concern the nuclear physicist.

The pre-eminent reference to the fundamental physical constants goes by the name 'CODATA', produced by least-squares adjustment of input measurement data every four years ($\cdots$, 2010, 2014, 2018, $\cdots$) and published two years later in the journal, *Reviews of Modern Physics*. There was also an interim evaluation in 2017 in preparation for the May 20, 2019 date when the SI Base Units were revised.

There are abundant sources for reading about the fundamental physical constants. Herein, we focus on aspects of this body of scholarship that are of interest to the nuclear physicist. Our interest is heavily dominated by mass-energy. This is often the determination of changes in mass-energy, but may use energy as a 'tag' for identifying a process such as a decay channel or study as a function of time of a radiation in a half-life determination. We have also noted above, the involvement of atomic hyperfine spectroscopy, and related studies using exotic atoms, i.e. the need to know the masses of particles such as muons and pions. Magnetic and electric fields are used to control and manipulate ions. Time 'tagging' of events is often of importance. Weak interaction processes connect to the strength with which nuclei couple to the weak field. As also noted earlier, intensities are measured by recording individual quantum events; and the task of detector efficiency calibration has many features unique to nuclear spectroscopy. From this perspective, the leading issue is what does a nuclear physicist need to know to practice their science. We follow with a few more brief remarks.

Remarkably, high precision in nuclear spectroscopic measurements is rarely needed. The reason is that the theory of nuclear structure rarely offers predictions that are expected to be of high precision. The only high-precision aspect of nuclear spectroscopy is that γ-ray cascades must obey the Rydberg–Ritz combination principle[4]. The highest precision needed for energies is in association with nuclear rotation of strongly deformed nuclei. The energies used to generate figure 2.12 involve precisions as high as a few electron volts in 50–100 keV. But there is no consensus regarding the significance of deviations from any simple empirical relationships. One use of high-precision γ-ray energies is the extraction of Doppler energy shifts to obtain lifetimes: small recoil velocities need high-precision energies. (We note at the extreme limit of small Doppler-induced γ-ray energy shifts is the so-called GRID technique conducted with the GAMS4 facility [9].)

[4] Sometimes, decay schemes depend on assignment of transitions on the basis of the Rydberg–Ritz combination principle. Great caution is needed in using this basis of assignments because of chance agreements. Use of coincidence information is advised wherever possible.

The key relationships needed by the nuclear physicist are those between the electron volt and the Joule, whence mass; and the atomic mass unit and the kilogram. With the new SI base units, 1 amu remains equal to 1/12 $M(^{12}C)$, exactly. We note that this could conceivably have been changed to a new amu equal to 1/28 $M(^{28}Si)$, in recognition of the role of ^{28}Si in the determination of Avogadro's number. But whether 1/12 $M(^{12}C)$, or 1/28 $M(^{28}Si)$, or $1/A\ M(^{A}X)$, nuclear binding energies render the choice arbitrary[5].

We conclude with some suggestions for further reading. An overview of metrology in the modern 'quantum' era is covered in [11–14].

7.3 Exercises

The relative challenge and time investment associated with the exercises is indicated from * to ***. Exercises marked with '**E**' require interactive access to figures or watching included video exercises and will only have such functionality in the e-book version.

7-1. Derive the relationship between energy and wavelength for photons: $E(\text{eV}) = 1240/\lambda$ (nm). [Hint: $E = h\nu$, $\lambda\nu = c$.] *

7-2. If the atomic mass scale should be changed from $u = M(^{12}C)/12$ to $u' = M(^{28}Si)/28$:
   (a) What would the percentage change be in going from u to u'? *
   (b) What is the physical origin of this difference? **

7-3. Suppose that the charge on the electron ($-e$) differed from that on the proton ($+e$) by a very small amount, say $-e - \delta e$: then hydrogen atoms would repel each other and gravity might not have been strong enough to produce stars and galaxies. Estimate an upper limit for $\delta e/e$. *

7-4. The length of the day is 86 400 s, except that it is never exactly this number!
   (a) By how much does the length of the day vary, day to day? *
   (b) There is a longterm systematic trend in the length of the day that necessitates 'leap' seconds. On what date did the last leap second get added/subtracted from the length of a day? *
   (c) How long has this been tracked? *

7-5. Everyone is familiar with leap years: they occur once every four years, and the determination of which years are leap years is whether or not the year is divisible by four. Except that is not the whole story.
   (a) Which will be the next year, divisible by four, that will not be a leap year? *
   (b) What is the general rule at this level of correction? *
   (c) How long before this corrected leap year rule, in turn, will need a correction? **

[5] The choice of the ^{12}C definition actually replaced a ^{16}O definition, by a process of evolving agreement, especially due to ^{12}C being a favored calibration mass. This occurred in the 1950s and 60s, especially under the persuasion of Josef Mattauch. The disparity had its origins in the atomic masses of chemists, which involved averages over isotopic abundances, and the nuclidic masses of physicists which were mono-isotopic. The 1961 nuclidic mass table of König, Mattauch, and Wapstra completed the shift to a ^{12}C-based definition [10].

7-6. The distance from the Earth to the Moon can be measured to a remarkable level of precision. At the level of $\pm X$ cm, what is X? **

7-7. Recently, the 'size' of the proton has needed revision.
   (a) How large was this change? **
   (b) What were the sources of the data that were in disagreement? (Which figures in this book illustrate the two types of data that were in disagreement?) ***

References

[1] Kessler E G Jr *et al.* 2001 The GAMS4 flat crystal facility *Nucl. Instrum. Methods Phys. Res.* A **457** 187

[2] Dewey M S, Kessler E G Jr, Deslattes R D, Börner H G, Jenstchel M, Doll C and Mutti P 2006 Precision measurement of the ^{29}Si, ^{33}S, and ^{36}Cl binding energies *Phys. Rev.* C **73** 044303

[3] Deslattes R D, Kessler E G Jr, Indelicato P, de Billy L, Lindroth E and Anton J 2003 X-ray transition energies: new approach to a comprehensive evaluation *Rev. Mod. Phys.* **75** 35

[4] Helmer R G, Hardy J C, Iacob V E, Sanchez-Vega M, Neilson R G and Nelson J 2003 The use of Monte Carlo calculations in the determination of a Ge detector efficiency curve *Nucl. Instrum. Methods Phys. Res.* A **511** 360

[5] Blank B *et al.* 2015 High-precision efficiency calibration of a high-purity co-axial germanium detector *Nucl. Instrum. Methods Phys. Res.* A **776** 34

[6] Schlamminger S 2018 *Redefining the Kilogram and Other SI Units Physics World Discovery* (Bristol: IOP Publishing)

[7] Robinson I A and Schlamminger S 2016 The watt or Kibble balance: a technique for implementing the new SI definition of the unit of mass *Metrologia* **53** A46

[8] Heyde K and Wood J L 2019 *Quantum Mechanics for Nuclear Structure, Volume 1: A primer* (Bristol: IOP Publishing)

[9] Börner H G and Jolie J 1993 Sub-picosecond lifetime measurements by gamma ray induced Doppler broadening *J. Phys. G Nucl. Part. Phys.* **19** 217

[10] König L A, Mattauch J H E and Wapstra A H 1961 Nuclidic mass table *Nucl. Phys.* **31** 18

[11] Cutler P H and Lucas A A 1983 *Quantum Metrology and Fundamental Physical Constants* (Berlin: Springer)

[12] Rainville S *et al.* 2005 A direct test of $E = mc^2$ *Nature* **438** 1096

[13] Mohr P J and Newell D B 2010 Resource letter FC-1: the physics of fundamental constants *Am. J. Phys.* **78** 338

[14] Fischer J and Ullrich J 2016 The new system of units *Nat. Phys.* **12** 4

Chapter 8

Epilogue

The material that we have assembled in the foregoing chapters has been selected with considerable thought, both with respect to what we view as an introduction or primer to nuclear data and with respect to what we have omitted as 'difficult to understand'. The core of the material selected provides the view that we use in our approach to a pedagogical communication of the nuclear many-body problem. Material that has been omitted lies at the research frontier with respect to how it should be organized, i.e. we lack a pedagogical view. In the following, we give a short account of what the present work achieves from our perspective, and where we are directing our attention in selecting topics in forthcoming books in the IOP Series in Nuclear Spectroscopy and Nuclear Structure.

The face of the nucleus that we have presented shows that it is a unique finite many-body quantum system that self organizes into a framework of shell structure, but with correlations. The correlations are of two types[1]: 'pairing' which dominates in singly closed shell nuclei, i.e. when only protons or neutrons are active; 'deformation' which dominates when both protons and neutrons are active. Even at doubly closed shells, deformation is observed at low energy in 'multi-particle-multi-hole' configurations.

Nuclear structure that is dominated by pairing, manifested in singly closed-shell nuclei, has the distinctive character evident in figures 4.18 and 4.19 and as explained in figure 4.20. There are only a small number of singly closed shell nuclei among those known. Nuclear structure that is dominated by deformation is widespread among known nuclei and that has been the main focus of this volume; however, this focus has been on 'strongly' deformed nuclei, i.e. nuclei that exhibit identifiable rotational bands. Thus, we have only provided brief details of 'weakly' deformed

[1] We forewarn the Reader that the issue of correlations extends beyond what is introduced herein, in important ways. Even at the foundational level of the independent-particle concept, there is a need to invoke interactions because < 70% of single-particle strength is observed in, e.g. knockout of protons by high-energy electrons, i.e. in (e,e′p) reactions. This is a more advanced subject that will be handled later in the Series.

nuclei. Our plan is to tackle weakly deformed nuclei (and other topics) in the next book in the Series. At a much deeper level, the reason that nuclei self organize into deformed shapes lacks a simple explanation at present.

With the continuation of the Series, we will be addressing issues of nuclear structure that are less well defined than those presented herein. This involves the need for more detailed spectroscopy; even a current lack of sufficient detail to arrive at a clear view. Notably, we will address issues such as low-energy vibrations in spherical and deformed nuclei, and triaxial deformation and other deformations. Consequently, the topics will move towards issues where a consensus is lacking and where the research frontier is active: this has been our intention from the outset. We note that the long history of the field of nuclear structure study has presented many structural ideas as unequivocal facts: and these ideas have either been disproven or are now being questioned.

In finer detail, the Series will provide a deeper look at experimental nuclear spectroscopic techniques and at nuclear models. Our intention is to sustain an 'open-ended' view so that emerging ideas and techniques can be accommodated. Progressively, the more advanced concepts in the quantum mechanics presented in *Quantum Mechanics for Nuclear Structure, Volume 1: A primer* and more in *Quantum Mechanics for Nuclear Structure, Volume 2: An intermediate level view* will be put to use. Individual experimental techniques will receive focused treatment in the Series.

As we progress, there are two components to this endeavor that we will aim to maintain: pedagogy and discovery. It would seem that for all of us, these two criteria deserve hard work.

www.ingramcontent.com/pod-product-compliance
Ingram Content Group UK Ltd.
Pitfield, Milton Keynes, MK11 3LW, UK
UKHW051853100926
8556IPUK00005B/600

9 780750 326728